BELOW THE EDGE
OF DARKNESS

BELOW

THE

EDGE

OF

DARKNESS

*A Memoir of Exploring Light
and Life in the Deep Sea*

*EDITH
WIDDER, PH.D.*

RANDOM HOUSE
New York

Published in the United States by Random House, an imprint and division of Penguin Random House LLC, New York.

RANDOM HOUSE and the HOUSE colophon are registered trademarks of Penguin Random House LLC.

Grateful acknowledgment is made to Counterpoint Press for permission to reprint "To Know the Dark" from *New Collected Poems* by Wendell Berry, copyright © 2012 by Wendell Berry. Reprinted by permission of Counterpoint Press.

LIBRARY OF CONGRESS CATALOGING-IN-PUBLICATION DATA
Names: Widder, Edith, author.
Title: Below the edge of darkness : a memoir of exploring light and life in the deep sea / Edith Widder, Ph.D.
Description: New York : Random House, [2021] | Includes bibliographical references and index.
Identifiers: LCCN 2020051565 (print) | LCCN 2020051566 (ebook) | ISBN 9780525509240 (Hardback : acid-free paper) | ISBN 9780525509257 (ebook)
Subjects: LCSH: Widder, Edith. | Marine scientists—United States—Biography. | Women marine biologists—United States—Biography. | Bioluminescence. | Underwater exploration.
Classification: LCC GC30.W54 A3 2021 (print) | LCC GC30.W54 (ebook) | DDC 551.46092 [B]—dc23
LC record available at https://lccn.loc.gov/2020051565
LC ebook record available at https://lccn.loc.gov/2020051566

Printed in the United States of America on acid-free paper

randomhousebooks.com

2 4 6 8 9 7 5 3 1

First Edition

Design by Fritz Metsch

For David

CONTENTS

A DIFFERENT LIGHT

There was a high-pitched whine coming from the starboard side of the sub. I leaned to my right, trying to pinpoint the source. It didn't necessarily sound alarming, but it *was* different, and one thing I have learned from diving in submersibles is to pay attention to *different*. Since this was the untethered single-person submersible Deep Rover, I was both pilot and crew, and there was no one to ask, "Do you hear that?" I was alone, just passing 350 feet beneath the ocean's surface, surrounded by water as far as the eye could see in every direction, and descending toward darkness. The whine, initially almost imperceptible above the whir of my scrubber fans, was growing louder and more concerning. Trying to identify the cause, I shifted around in my seat, a padded pilot's chair in the center of the five-foot clear acrylic sphere. I bent at the waist, contorting to bring my right ear almost level with the instruments in the armrest. As I did, my stockinged feet slid down the inside of the sphere, into something you never want to encounter inside a submersible: seawater. Lots of it.

This was bad. Abject terror was the appropriate response, which I certainly felt in spades. Fortunately, I was able to maintain just enough of my faculties to begin taking action to save myself. Locating the source was step one. No problem: The water was flowing in through an open valve below my seat on the starboard side. Stopping it was step two. Big problem: *The valve handle was missing!* There was only the valve stem, which I couldn't possibly turn without leverage. The water kept streaming in through the small orifice, the whine's rising pitch providing an audible measure of the sub getting heavier and sinking faster as the water rose. I blew my ballast tanks and jammed on my vertical thrusters while my mind raced. *Is it too late? Am I already past the point of no return?*

The fact that I am writing this proves, of course, that I wasn't too late. I made it to the surface and was pulled to safety, but it was definitely more harrowing than I would have liked, and I can't deny that the memory lingers.[*] Over the course of my career as a marine scientist, I have made hundreds of dives in submersibles, which means there have been other bad moments—not many, but enough. This was not the worst,[†] but it did happen early in my career and could have cost me my life, so why, you might ask, do I keep doing it? Honestly, it never crossed my mind to stop.

I became hopelessly addicted in 1984, on my first dive using a metal diving suit called Wasp. That initial spine-tingling exposure to deep-sea bioluminescence occurred during an evening dive off the coast of Santa Barbara. I was dangling on the end of a cable at what at the time seemed an incomprehensible depth—800 feet—where the pressure outside my protective metal shell was 355 pounds per square inch (24 atmospheres). I was there to explore and learn about life in the largest living space on the planet, the ocean's midwater. I hoped I would see bioluminescence, which is why I turned out the lights. I wasn't disappointed. In fact, I was so dazzled, the experience changed the course of my career.

---

[*] Notwithstanding, I have found the memory useful when dealing with stressful situations. *Could be worse—could be seawater in the sub* has proven to be a serviceable mantra for restoring perspective.

[†] Just a very close second.

*   *   *

THE FIRST MAJOR scientific collecting cruise of marine life in the diverse Sea of Cortez was organized by Ed Ricketts and John Steinbeck in 1940. Their explorations and scientific discoveries were published in *The Log from the Sea of Cortez*, which was written by both of them but is generally ascribed solely to Steinbeck. Their vessel was a chartered purse seiner called the *Western Flyer*. During their expedition, Steinbeck and Ricketts and the rest of their crew made most of their collections at low tides. With their bent-over posture and slow head-scanning movements, they would inevitably draw questions from the locals:

"What did you lose?"

"Nothing."

"Then what do you search for?"

This line has always made me laugh, not least because I've been asked this question or something close to it many times in my forty-year career of exploring the ocean. And the truth is, I've asked it of myself more than a few times.

Encounters with bioluminescence at scale can make that question seem quaint. The open ocean is a fantastically strange and wonderful place. In this world without apparent hiding places, the game of hide-and-seek is played out on a daily basis with life-and-death consequences. One successful survival strategy is to hide in the depths during the day, below what we call the edge of darkness, and come up only to feed in food-rich waters at night, as the edge of darkness makes its way toward the surface. This is such a common solution to the problem of no hiding places that it is responsible for the most massive animal migration pattern on our planet.

Vertical migration happens every day, in every ocean, and the masses of ascending animals form a layer so dense that more than one ship's captain, scanning with sonar at sunset, has been fooled into believing they were about to run aground. Because so many of the ocean's inhabitants have adopted this survival strategy, these migrants spend most of their lives in near darkness. To compensate, almost all of them make their own light.

Drag a net behind a ship almost anywhere in the ocean below the edge of darkness, and most of the animals you bring up in that net will make light. Given the volume of the open ocean and the vast watery realm between the ocean's surface and its bottom, which constitutes the largest ecosystem on the planet, we're talking about a world teeming with light makers. To put this in perspective, if most of the animals in the ocean are bioluminescent (from single-celled bacteria to colossal squid), then a majority of the creatures on the planet are communicating using language-of-light dialects that we don't comprehend.

Bioluminescence's power to captivate is evident in the descriptions of anyone so fortunate as to experience it firsthand. The adjective heard most often is *magical.* The pure magic of living light hearkens back to childhood fantasies of secret grottos, wizards' caves, and unicorn haunts, where the mushrooms in fairy rings glow with cold green fire and a wave of the hand sends multicolored sparks streaming from fingertips. Real-world encounters with such enchantments manifest as children chasing fireflies on warm summer nights, lovers strolling a beach hand in hand with the Milky Way overhead while sprinklings of sea sparkle gild their footprints in the sand, and kayakers on a moonless night creating luminous blue explosions and sprays of liquid light with each dip and arc of their paddles. For these lucky few, bioluminescence is not one of nature's obscure and little-known oddities; it is one of their most precious and lasting memories.

That there are even more spectacular light shows in the inky depths of the ocean was little known when I made that first deep dive. There was plenty of scientific evidence of bioluminescence's presence in the form of light-meter measurements and net-captured creatures (mostly dead) studded with light organs, but there were very few direct observations, and none that captured anything like the spectacle I witnessed. This was a light extravaganza unlike anything I could have imagined. Afterwards, when asked to describe what I had seen, I blurted, "It's like the Fourth of July down there!"—a highly nonscientific description that got quoted in our local newspaper and for which I took a healthy dose of ribbing from my colleagues. However, I have had many opportunities since then to take people on submersible dives, and I have lost count of the number of times I have heard them described as "like the Fourth of July!"

Fireworks are an extraordinary art form—light painted onto the black canvas of a night sky. Each brushstroke is a splash of photons defined not just by color and contour but also by movement through space and time. Transience is what saves these spectacles from the kitschiness of a black velvet painting, as each burst of light morphs from moment to moment, rocketing up, blooming out, cascading down—incandescent for mere moments before disappearing into nothingness. There are identifiable components to the displays, each unlike any that has come before or will come again—but still recognizable as chrysanthemums, palms, Roman candles, or horsetails. Their frequent repetition produces a visual echo—variations on a theme, like literal light jazz—in the pictorial rather than the teeth-grinding musical sense.

I once purchased a book of photographs for a single image. It was a photo, captured by *Life* magazine's Gjon Mili, of Pablo Picasso drawing with light. The story goes that Mili had been experimenting with light painting by attaching tiny lights to ice skates and then photographing the skaters jumping in the dark. Picasso, who originally had been less than thrilled at the prospect of being photographed for the famous magazine, became intrigued when Mili showed him these experiments. Picasso was so tantalized by the possibilities that he ended up posing for five sessions, in which

he produced thirty light paintings. The one that prompted my book*
purchase is the best known. Taken in Picasso's pottery studio, it is
a black-and-white and shows a flash image of the famous artist in a
semi-crouched position, staring directly into the camera, holding
the small illuminated lightbulb that he then used to draw a centaur
in the dark while the camera shutter remained open. In the photograph, the centaur hangs suspended in space between the artist
and the camera—an ephemeral phantom—disappearing even as it
was being created, but preserved in its entirety on photographic
film.

Just as fireworks are a kind of light painting, so is bioluminescence. But instead of the product of incandescence and human ingenuity, bioluminescence is cold chemical light, a consequence of
millions of years of evolution resulting in fantastic light-emitting
creatures with evocative names like crystal jelly, cockeyed squid,
bearded seadevil, shining tubeshoulder, stoplight fish, and velvet
belly lantern shark. Their bodies are adorned with all manner of
light-emitting structures—nozzles that spew liquid blue flame, incredibly complex light organs that look like flesh-encrusting jewels
but behave optically like eyes that *emit* light instead of collecting it,
and absurdly elaborate glowing appendages that resemble abstract
sculptures of alien life-forms from the planet Zork.

On that first dive in Wasp, I had no idea who was producing the
dazzling light I was witnessing. It was so extreme and otherworldly
that the only description I had for it was my lame Fourth of July comment, but of course that didn't do it justice.

It's true that there were evanescent sprays and swirls and squirts
of flashiness glittering in the darkness that resembled the "light
jazz" of a fireworks display, but they weren't every color of the rainbow. Instead, it was a mixture of the most brilliant blues ever to

---

* John Loengard, *Life Classic Photographs: A Personal Interpretation* (Boston: Little,
Brown, 1988).

grace an artist's palette—azure, cobalt, cerulean, lapis, neon—supernatural hues, emitted rather than reflected light.

And, unlike on Independence Day, I wasn't a passive observer from afar. I sat at the very center of this show. In fact, I was *part* of it, because, as I quickly discovered, any movement I made was a trigger. Even very slight motions would stimulate spreading spheres of aquamarine sparkles and flashes. And if I activated the Wasp's thrusters, glorious streams of light would erupt from the propellers in vortices of shimmering cobalt-blue liquid mixed with what looked like glowing embers swirling up off a campfire as you toss on a new log, only these were icy blue. But if I attempted to see who was making the light by turning on my floodlights, I saw no one; the space was devoid of any recognizable life-forms.

All of this bioluminescence was the product of life, but in this case, it was apparently generated by creatures too small or too transparent to show up in my floodlights. I knew how much energy it takes for life to generate light* (a lot!), so I was keenly aware that this was no trivial phenomenon. Energy expended on this scale has enormous significance, but on that very first dive, I had no idea what that significance could actually be. That tantalizing mystery hooked me then and has kept me going back again and again to try to better understand the role living light plays in the vast oceanic realm that is so frequently described, incorrectly, as a world of eternal darkness.

Marine biology may be one of the careers kids most say they want to get into. But what is the point of this labor-intensive, not undangerous, sometimes expensive, and vastly underappreciated science? Just what are we searching for, and, most important, what have we found?

---

* In a highly efficient reaction, one photon is equivalent to the hydrolysis of about 6 ATPs (the energy currency of life). In a less efficient reaction, it may be ten times that. So a single-celled microscopic dinoflagellate (a.k.a. sea sparkle) that emits about 10^{10} photons per flash is using up that currency at a rate of between sixty billion and six hundred billion ATPs per flash.

The desire to understand the natural world resides in all of us. Exploring and sharing knowledge about how the world works is foundational to our survival. In primitive humans, the drive to uncover the secrets of nature advanced essential life skills such as finding sustenance and shelter and determining what animals were deadly and what food was safe to eat. In modern times, our drive to explore has led us to extraordinary discoveries, ingenious innovations, and some of our most fantastic achievements. So how is it possible that we have yet to explore what constitutes the largest living space on our planet—the deep ocean?

Inaccessibility in the form of crushing pressures is certainly a challenge, but it is one we have met and overcome. Cost, too, poses a stumbling block, but it can't be the only reason for the dearth of exploration, when we have spent trillions to land on the moon and Mars. Rather, the greatest obstacle may be the widely held misperception that there is nothing left to discover on Earth. One of the rationales sometimes given for space exploration is that everything on this planet has been climbed, crossed, and spelunked.

The truth is that the staggering reaches of the ocean that remain unseen exceed, by many times, all the territory *ever* explored. We seem to be in a catch-22 scenario where we haven't explored the deep ocean because we don't appreciate what a remarkable, mysterious, and wondrous place it is, and we don't know what an astonishing place it is because we haven't explored it. What makes this situation all the more untenable is that we are managing to destroy the ocean before we even know what's in it.

In the course of human history, our pattern has been exploration followed by exploitation, but in the ocean we have managed to reverse the order—massively exploiting the ocean's resources before exploring what's actually there. In the past sixty years, we have altered the ocean more than in all of the preceding two hundred thousand years of human existence. We have stripped it of its big fish, using nets so enormous they could hold a dozen jumbo jets, and we have deployed hook-studded longlines as much as a hundred miles long that target big predators like tuna, swordfish, and halibut but also snag turtles, dolphins, and even diving seabirds. We have harpooned whales, driving these magnificent, intelligent creatures to near extinction. Our bottom trawlers drag vast weighted nets across the seafloor, turning exquisite undersea gardens full of living beings into rubble piles that won't sustain life again for hundreds of years.

At the same time that we are pulling out every last fish, shrimp, and squid, we are filling the ocean with our plastics, trash, and toxins. It is estimated that by the year 2050 the ocean will contain more plastic by weight than fish. Traces of radioactive waste, PCBs, mercury, and chlorofluorocarbons are detectable in the depths of the ocean and for all we know are already impacting the life down there.

We are poised on the brink of massive destruction of oceanic ecosystems—systems that are vital cogs and wheels in the extraordinary machinery of all life on planet Earth. The enormity of the ocean has managed to protect us from ourselves for a long time, but our exploding population and mismanagement of resources are beginning to overwhelm even its phenomenal buffering capacity.

For the past couple of centuries, the ocean has been working mightily to keep the all-important carbon cycle in balance by absorbing much of the vast quantity of carbon dioxide we have been releasing through the burning of ancient sequestered carbon, that is, fossil fuels. In the process, the ocean has been acidifying, because when carbon dioxide dissolves in seawater, carbonic acid is formed. Let's pause for a moment to grasp the gravity of that statement. We are changing the chemistry of a staggering quantity of water—a volume of three hundred million cubic miles! This change is beginning to alter the food web, making it difficult for keystone species like corals, shellfish, and sea butterflies* to construct their skeletons and shells.

---

*Also known as pteropods, they are an important food source for many fish, whales, and seabirds.

The ocean has also been absorbing much of the extra heat that has been accumulating because increasing concentrations of greenhouse gases, like carbon dioxide and methane, in our atmosphere are preventing heat from radiating from the Earth's surface and escaping into space as readily as it used to. That heat buildup has many worrisome implications.

Warming waters and melting ice are potentially altering the flow of the great rivers in the sea like the Gulf Stream. More than sixty miles wide and a half-mile deep, the Gulf Stream transports a volume of water that exceeds that of all the rivers in the world by twenty-five times, carrying warm water from the equator along the east coast of North America and across the great Atlantic to northwestern Europe. This is one of several massive rivers that flow throughout the ocean in a complex pattern known as the Great Ocean Conveyor Belt. Arising from density differences between colder, salty water masses and warmer, fresher water masses, the conveyor belt has major impacts upon weather as it moves heat around the globe. Changes in these flow patterns may already be linked to increasing droughts, floods, hurricanes, and wildfires; destabilization of agriculture and fisheries; and untold increases in human suffering.

The list goes on and on: plastics pollution, chemical pollution, nutrient pollution, overfishing, invasive species, bottom trawling, seafloor mining, deep-sea drilling, destruction of estuaries and wetlands, loss of coral reefs and sea ice habitat, and—one of my personal favorites, from the darkest recesses of my anxiety closet—the thawing of subsea and land-based permafrost,* causing the release of substantial quantities of greenhouse gases as microbes go to work on the vast amounts of organic matter therein. This could lead to a tipping point, sometimes referred to as the permafrost bomb, akin to the kind of feedback loop I was facing when water was coming into the Deep Rover. Once we reach that point, there's no going back.

---

*A thick subsurface layer of organic-rich frozen soil.

There have been endless books, science papers, magazine articles, documentaries, and social media posts detailing the doom and gloom—to little avail. We are still essentially fiddling while the planet burns. There are a couple of psychological components to this.

One is illustrated by the boiling-frog story that claims if a frog is dropped directly into boiling water it will have enough sense to jump out, but if it is placed in warmish water that is heated slowly, it will hang around until it is stewed to death. Personally, I suspect frogs are smarter than that, and I certainly want to believe humans are.

The other rationale for our inaction is the simple fact that the swelling drumbeat of decimation engenders such a sense of helplessness that people want to plug their ears and cover their eyes. But the beat goes on, growing ever louder, with the hope that if we just point out how truly dire the situation is becoming, the appropriate checks and balances will be brought to bear, like taxing carbon and switching from burning fossil fuels to alternatives like solar, wind, hydroelectric, geothermal, and nuclear energy. Clearly that's not happening, at least not on the time scale we need it to happen. In fact, in many cases the drumbeat seems to be having the opposite effect.

It has been said that Martin Luther King, Jr., did not mobilize the civil rights movement by preaching, "I have a nightmare."* Nonetheless, that's what many on the environmental front lines are doing. We need a different outlook, one that focuses on our strengths rather than our weaknesses. Exploration has always been the key to our survival, which is why I believe we need explorers now more than ever. Explorers are, by necessity, optimists who have to see beyond imagined limits to find a way forward. They push past the scary monsters at the edge of the map and have the persistence needed to

---

* I'm not sure who first put forth this notion. I have heard a number of people use it, but I think it originated with statistician Nic Marks in his wonderful 2010 TED talk "The Happy Planet Index."

pursue solutions in the face of seemingly impossible odds. Their tenacity often arises less out of an abundance of courage than from an abiding curiosity.

Throughout our lives, unexplored places that promise fantastic discoveries entice our imaginations. The stories that mesmerize us when we're young revolve around discovering portals to other worlds: unearthing the entrance to an ancient tomb full of treasures, discovering a hidden gate that leads to a secret garden, or following a rabbit down a hole to a fantastic wonderland. Mysterious, unexplored places draw us. What few people realize is that most of our planet remains mysterious and unexplored. The ocean's depths hold some of the most fantastic secrets about life on Earth and answers to questions we haven't even thought to ask.

I have a quote pinned next to my office window, overlooking one of Florida's ocean inlets: "The world will not perish for want of wonders, but for want of wonder."* Our survival on this planet depends on fostering a greater sense of connection to the living world, and wonderment is key to forging that link. I have long believed that bioluminescence provides a means to reveal the wonder in this unseen world to a public that is alarmingly unaware and, thus, largely indifferent to what makes life possible on our planet. I believe it is a light capable of exciting the imagination and firing the inborn curiosity that defines the core of what it means to be human. I hope it can fire the imaginations of the next generation of explorers and in so doing provide a beacon of hope for the future of life on Earth.

---

* Usually attributed to the poet and theologian G. K. Chesterton, but sometimes to the scientist J. B. S. Haldane.

PART I

DEEP SEEING

The only true voyage of discovery,
the only fountain of Eternal Youth,
would be not to visit strange lands but to possess other eyes.

—MARCEL PROUST

Chapter 1

SEEING

Light is *what*, exactly? For something that has no mass, it sure carries a lot of existential weight. It is both an energy source and an information carrier. It can be injurious and it can be healing. It is one thing that can manifest as two things—a wave in the future and a particle in the past. In a vacuum, it travels at the maximum speed allowed by the universe, and it does so without ever decaying. It gives up its energy only when it interacts with other particles, like those that make up the visual pigments in our eyes, and it is through these interactions that we interpret the world around us.

For most life-forms on Earth, light is the paramount stimulus that makes life as we know it possible. Green plants harness energy from light to synthesize sugar from carbon dioxide and water. In the process, oxygen is generated as a by-product. As magic tricks go, forming food and breathable air from what seems like nothing is hard to beat. Still, it's not especially flashy. Creating dazzling light from food and air, however, is *very* flashy. That's the magic of bioluminescence. Of course, to appreciate that particular alchemy, you need something equally miraculous: vision.

Being able to see provides a huge advantage in the game of life; it is for this evolutionary reason that 95 percent of all animal species on Earth have eyes. These range from microscopic, such as some single-celled algae that have an eye no bigger than one-tenth the diameter of a human hair, to giant squid with an eye the size of your head.* The different ways that such disparate eyes see the world reveal much about the biological needs of their owners. In fact, figuring out what different eyes are best adapted to see is such a valuable tool for probing the nature of life that it has become a whole field of study called visual ecology.

If you compare the life of a giant squid inhabiting the deep sea with that of microscopic plankton living in sunlit surface waters, the difference in eye size makes sense: a giant eye collects many more photons than a tiny one and is therefore better adapted for living in a dim light environment. But what about another deep-sea inhabitant, the cockeyed squid? Its name derives from its mismatched eyes: the left eye is giant and bulging and directed *upward* toward the sunlight, while the right eye is smaller, recessed, and aimed *downward* into the inky depths. This seemingly makes no sense—until you learn that bioluminescent light organs encircle the small eye. While the large eye hunts overhead for dim, distant silhouettes of prey against a dark, lead-gray background, the bottom eye can use its built-in flashlights to illuminate more proximate prey. Clearly, to understand the visual ecology of the largest living space on Earth, one needs to appreciate the nature and function of bioluminescence alongside the nature and function of eyes.

It is inevitable that when we try to figure out what different animals see, we relate it to what *we* can see. That is a major challenge in the deep ocean, though, where our very presence alters the visual environment. It's difficult to envision a place you are unable to observe in its natural state. Our eyes are adapted for a much brighter existence, which means that when we explore darkness, we must bring artificial lights so intense that to visual systems adapted to the

---

* In metric units, that's ten microns and thirty centimeters, respectively; in American units, it's one-tenth the thickness of a dollar bill for the former and one-fifth as tall as Danny DeVito for the latter.

deep sea, they are probably as bright as looking directly into the sun. Since it is such a challenge to observe animals in this realm without disturbing them, sometimes the best way to gain insight into their lives is to learn as much as possible *about* their eyes.

The most important questions to ask about eyes are: What information do they accept, and what do they exclude? All eyes act as filters, allowing in only data streams about the outside world that optimize their owner's chances for survival. Anything that doesn't serve that purpose falls under the banner of *too much information*. Spending time and energy on producing ultraviolet receptors, for example, and processing and interpreting their output is counterproductive if UV light plays no useful role in detecting vital stuff like food, mates, or predators.

Thinking about eyes and what they do and don't see is a mind-stretching exercise. We are blind to so many things in our world—some because of biological constraints, and many more because we simply don't know how to look. Environmentalist Rachel Carson once said, "One way to open your eyes to unnoticed beauty is to ask yourself, *What if I had never seen this before? What if I knew I would never see it again?*" An even better way to achieve heightened visual awareness is to lose sight and then regain it. As Joni Mitchell sang, "Don't it always seem to go, that you don't know what you've got till it's gone?"

THAT JONI MITCHELL song, "Big Yellow Taxi," was released my first year in college. I started at Tufts University in the fall of 1969 as a biology major, with the aim of becoming a marine biologist. But before I had completed my first semester, it was clear that goal would be unattainable without medical intervention. During my precollege physical, I mentioned a pain I'd been having down the back of my left leg. Since I was pretty active—a skier and skater in the winter and a water skier in the summer—I figured I must have pulled a muscle. X-rays revealed otherwise: My back was broken. The doctor illustrated the extent of the break by making two fists with his hands, stacking one on top of the other, and then sliding the top one half-way off the bottom one. The slippage was pinching a nerve going down my left leg, causing the intense and persistent pain I felt whenever I sat down.

I'm pretty sure I know when I broke it. I spent a lot of my child-hood climbing up into and jumping out of trees in our leafy subur-ban neighborhood just outside Boston. My favorite tree was an old misshapen willow down by the pond* near our house. Its trunk ramped up at a forty-five-degree angle away from the water and then branched into two large horizontal limbs, each with thick vertical branches that created separate "rooms" that made it the perfect pi-rate ship, tree house, or castle. The limbs were about seven feet off the ground: a comfortable jump that I made hundreds of times with ease. But I remember one Sunday, when I was eight or nine years old and dressed for Sunday school in some stupid frilly dress I hated, the jump didn't go as planned.

When we came back from church, I couldn't change into my be-loved jeans, because we were going someplace fancy later, but I was allowed to be outside until it was time to go, as long as I promised to stay clean. I wandered off to climb my favorite tree, but when I went to jump down, I remembered my promise and landed in a way that protected my dress instead of myself. A searing pain ripped through my back—like nothing I'd felt before. But it didn't last long, and I shrugged it off.

Until that college physical, I thought low back pain was some-thing everybody had. I couldn't remember a time without it. By my first semester at Tufts, it was so bad that I couldn't stand for any ex-tended period, and sitting was equally miserable because of the pain

---

*Actually, at 103 acres this is more of a lake than a pond, but in New England "pond" is often used in reference to some of its lakes' origins as "kettle hole ponds," depressions left by receding glaciers.

in my leg. The only way I could do homework was by lying flat on my back with a pillow under my knees. This was not conducive to good study habits, as I would often fall asleep and bonk myself in the face with whatever tome I was attempting to wade through—a very effective form of negative conditioning. When it became clear I couldn't go on like this, a spinal fusion was scheduled for the beginning of February.

ACCORDING TO THE Urban Dictionary, "crumping" is a slang medical term indicating that a patient's condition is rapidly worsening. See also: "circling the drain."

I crumped; not during the spinal fusion, which went fine, but afterwards, in the recovery room. I went from okay to *Oh shit* in a New York minute,* flipping around in the bed like a fish on a dock while hemorrhaging nearly everywhere. I had a blood disorder called disseminated intravascular coagulation (DIC). The cause is unknown, but it's often associated with major trauma and manifests as overactive clotting factors in the blood, causing clots to form in the small blood vessels of the body, blocking blood flow to vital organs. In extreme cases, the clotting factors and platelets are consumed to such a degree that severe bleeding ensues. The result, in my case, was that I wasn't just bleeding into my surgical sites but also into my lungs, depriving me of air, hence the fish-out-of-water imitation.

Two factors conspired to allow me to survive this medical Armageddon. In fact, I was the first person *ever* to survive it at Mount Auburn Hospital. The first was that my orthopedic surgeon had recently attended an American Medical Association conference on DIC, which allowed him to recognize the symptoms. Usually, a doctor who sees his patient hemorrhaging will administer coagulants to stop the bleeding, but that just leads to more clotting in the small

---

*The time between the traffic light turning green and the cab behind you starting to honk.

blood vessels and increased likelihood of organ failure. Instead, my surgeon knew to give me the anticoagulant heparin, thereby averting organ failure but greatly exacerbating the bleeding problem.

The second lucky break was that the famous Harken* "chest team" happened to be at Mount Auburn that day. The chest team's first order of business was to start my heart, which had stopped. Next, they needed to clear the blood from my lungs, which required flipping me from my back onto my side. Then the whole process had to be repeated as my heart stopped again and my lungs refilled with blood. In total, I had to be resuscitated three times.

I had *three* resuscitations, but only *one* near-death experience. My NDE involved a classic out-of-body experience, or OBE, in which I was looking down at myself from above. There was another consciousness there with me—no physical presence, just a noncorporeal being—and we were trying to make a decision about the outcome of the current situation. I remember feeling completely neutral, fine either way. As someone who has personally experienced an NDE, I understand the temptation to assign a spiritual explanation. It certainly didn't *feel* like a dream; it felt real. As a scientist, though, I have learned to be comfortable with ambiguity, and I choose to keep an open mind on the subject.

The most intriguing thing to me about NDEs is their commonalities. Sometime later, when I read Elisabeth Kübler-Ross's *On Death and Dying*, I discovered that my experience was not unique. A frequently cited feature of NDEs is the sense of peace felt both during and immediately after the experience. All the usual brain chatter related to time and tasks is silenced. After my NDE, I felt present,

---

*Dwight Harken (1910–1993), known as the father of heart surgery, was chief of thoracic surgery at Peter Bent Brigham Hospital in Boston and Mount Auburn Hospital in Cambridge from 1948 to 1970. That appointment meant that his and his team's being at Mount Auburn that day was somewhat less of a miracle than I was originally led to believe—unless you factor in the traffic between Boston and Cambridge.

fully in the moment, in a way that I had never experienced before and haven't since. I was the opposite of solitary and separate, but rather at one with, and connected to, everything. As a result, I was unfazed by the fact that when I regained consciousness I was a total mess—a veritable pincushion with tubes and wires running everywhere, on a breathing machine with a tube down my throat that prevented me from speaking. And I was blind.

It's strange, in retrospect, that none of this bothered me in the slightest. When the doctors and my parents explained to me what had happened, it all sounded familiar and understandable. Somehow, I was at peace with it.* In fact, I had accepted my blindness so completely that I didn't tell anyone I couldn't see until several days after the surgery, and when I did say it out loud, it still didn't seem that important.

That sense of peace lasted for a week in the ICU and persisted for a few more days after I was moved to the pediatric ward in the Wyman wing of the hospital. At eighteen years old, I no longer qualified for that ward, but it was the only floor with an available observation room where the nurses could keep an eye on high-risk patients. I would remain there for the next four months.

Once out of the ICU, I was allowed visitors other than family. Friends brought flowers, including one especially beautiful bouquet of roses that everyone commented on when they entered my room. On one such day, someone mentioned the beautiful "yellow" roses. *Wait. WHAT? Yellow?* I felt a shock of adrenaline as I abruptly came to grips with the extent of my blindness. It was as if someone had slapped me awake. I thought the roses were red, but that's merely the color I *assumed* they were. My analytical brain jump-started and I quickly ran through an assessment of what I could actually see.

---

* Nerd note: I wonder if NDEs might be an evolutionary adaptation allowing rational thought and calm in such extreme, life-threatening circumstances, where adrenaline and panic might otherwise result in further, self-inflicted damage.

The answer was *not much*. I couldn't see the roses; I had just imagined them to look like stereotypical red roses. I couldn't see the door to my room, but I had mentally sketched it in based on the direction of sounds as people entered and exited. I couldn't even see my own hand held up in front of my face; I just knew it was there because I could feel myself holding it up.

I thought I was seeing the people I knew who visited me, but now I realized that, too, was a fabrication of my imagination, because I couldn't remember the faces of any of the parade of doctors, nurses, and technicians coming in and out of my room every day—faces I hadn't met before I crumped. There were brief glimpses of light and shadow, but no real visual information.

THE DISPARITY BETWEEN light and dark is striking, so much so that it figures significantly in many creation stories. Out of darkness and nothingness come light and somethingness. We associate darkness with chaos and light with a sense of order, but that order is contingent upon the capacity to see and make sense of what the light reveals. Being able to detect light and dark is certainly better than nothing, but it's a far cry from the phenomenal advantages afforded by true eyesight.

Vision occurs in three phases. Phase one happens in the eye, which, like a camera, focuses an image of the world onto a light-sensitive surface. But where a camera uses film, our eyes have retinas, each composed of 126 million light-sensitive cells called photoreceptors. Phase two converts light energy into electrical signals that travel through a sequential series of neurons to the brain. And in phase three, the brain interprets those electrical signals, forming a mental image. That's the purpose of vision: to create a link between the physical world and that all-important central processor, the brain. The survival advantage that results from being able to respond appropriately to threats and opportunities requires much more than just recognizing images. Objects must be identified from multiple perspectives, distances gauged, and motion and trajectories calculated. And all of this needs to be possible even when the viewer is on the move. The fact that when you tilt your head, the whole world doesn't appear to tilt is just one small example of the enormous processing power your brain affords.

How our visual system handles information reveals a lot about our perception biases. For example, we are far more interested in detecting contrast—differences in intensity—than we are in absolute intensities. There are some fantastic optical illusions that demonstrate this. One of my favorites is what's known as the checker shadow illusion, which is a shades-of-gray graphic of a checkerboard with a large cylinder sitting on one corner. The scene is side-lit, creating a shadow of the cylinder across the board. It looks perfectly reasonable until someone tells you that one of the "white" squares inside the shadow is the exact same shade of gray as one of the "black" squares outside the shadow. Nobody believes it. The only way to convince yourself that it's true is by somehow eliminating the surrounding cues that your brain is using to interpret the scene. For example, you can cut out the two squares and hold them side by side. They really are the same, but when you view them as part of the whole scene, your brain adjusts your perception by pumping up the apparent brightness of the one in shadow, thereby enhancing the contrast.

At the neural level, this contrast bias can be detected by measuring the electrical activity of ganglion cells in the retina. A ganglion cell, which receives input from a small patch of photoreceptors, gives a much bigger electrical response to a small spot of light shining on the center of that patch than to diffuse illumination across the whole area. And farther up the chain, in the optical processing center of the brain, neurons are so enamored of contrast that they are essentially unresponsive to uniform illumination.

Individual cells are also biased in favor of detecting movement. You can demonstrate this if you place someone before a motionless scene, secure their head with a skull clamp, and anesthetize their eye muscles.* With no motion in their visual field, they become essentially blind. You can sometimes approximate this effect when you "zone out" in a relaxed state, staring fixedly at a point without moving or blinking, causing your peripheral vision to "white out."

For the brain to make sense of the three-dimensional world based on the upside-down, two-dimensional image projected on the retina, it must deal with an astonishing amount of ambiguity. In fact, for any retinal image there is actually an infinite number of possible three-dimensional forms that might have generated it. Consequently, the brain is constantly having to extrapolate information from sparse input.

One of the early investigators working on how the brain interprets sensory input was Karl Lashley (1890–1958), an American scientist best known for his research on learning and memory. Lashley suffered from migraines so intense that they were both figuratively and literally blinding. During one such attack, he made the intriguing observation that, while the area of complete blindness in the center of his visual field obscured the head of the colleague he was looking at, the vertical stripes on the wallpaper behind him appeared to pass right through where his colleague's head should be. If the man's head had been visible, it would have masked the stripes, but in the absence of actual input from that part of the visual field, Lashley's brain filled in the most likely image based on the surrounding visual field. That's a heck of a trick, but it's a trick that should give you pause and raise questions about how much of what you see is biased by what you *expect* to see.

MY VISION LOSS was in phase one. I had hemorrhaged into the large space between where light is focused at the front of the eye

---

* Preferably with their permission.

(through the cornea and lens) and the retina, at the back of the eye. This region, known as the vitreous chamber, is filled with a transparent, colorless, gel-like substance called the vitreous humor, which functions primarily to hold the spherical shape of the eye, transmitting light unimpeded with a sharp focus on the retina.

The process of vision depends on being able to make comparisons between light rays coming from different directions. Two factors determine performance: sensitivity and resolution. Sensitivity is all about the number of photons it takes to generate a recognizable signal. Resolution is akin to the number of pixels that make up a photograph—specifically, it is the number of photoreceptors per area of retina in combination with the clarity of the image created by the eye's optics.

The blood that had leaked into my eyes was absorbing and scattering the light, thereby impeding both sensitivity and resolution. Through my right eye I could make out a little light, but the left eye was a full eight-ball hemorrhage, letting virtually nothing in. My visual world was swirling darkness with occasional glimpses of meaningless light.

The doctors were distressingly noncommittal about my prognosis. There was so much blood in the vitreous humors that they couldn't see the retinas, which meant they couldn't be sure they hadn't detached. All they would say was that the body would clear blood from the vitreous chamber, but it can take months. There was nothing to be done but wait and hope for the best. In the interim, I was simply trying to hang on tight to the roller coaster of recovering from major trauma. It's a freakish, terrifying ride, where the lows always seem so much lower than the highs.

Then they discovered a massive infection in my surgical site, and cleaning out that pocket of infection required emergency surgery. This was done with only local anesthetic, because of concerns that general anesthesia may have contributed to the DIC. It was agonizing* and terrifying and had to be repeated every other day for a month. At the same time, I was receiving antibiotics intravenously at maximum dosages that burned through my veins and required lots of poking and prodding with IV needles to reestablish the drip. I had a variety of interesting adverse reactions to the different antibiotics, like rashes and boils. Then, just as those woes subsided, I came down with serum hepatitis, thanks to the twenty-three blood transfusions I had received. This involved intense liver pain, lots of vomiting, and a general yellow tinge that looked like a cheap spray-on tan.

And somewhere in the midst of all this, my doctor informed me that my spinal fusion was no good. The "glue" for the fusion was supposed to be made of bone chips taken from my hip and packed around my vertebrae. However, as my doctor regretfully explained, all the flipping around I had done on the recovery room table had sent the bone chips flying: X-rays revealed there was virtually nothing left to grow into a fusion. I was devastated. It was inconceivable that I had suffered through all this for nothing.

Everything that was happening to me seemed so out of my control that I came to realize that my only power came from my mindset. Small things mattered a lot. The big picture was far too scary to contemplate. It was like being trapped on the side of a cliff face with no clue how long I would have to climb to reach safety. "Don't look down" is good advice if doing so will give you vertigo; "don't look up" is also sage counsel if there is no end in sight. The only thing I could do, I realized, was concentrate on finding that next handhold.

That ability to shift mental focus became my key to coping with what seemed like an endless series of setbacks. Instead of looking

---

* Either they hadn't administered enough local or my brain was interpreting pressure as pain. Much later, one of the surgical nurses who was on duty that night admitted that she had never seen anything like what they saw when they opened me up. After they cleared all the necrotic tissue away, she said, there was a cavity big enough to stick a man's fist into.

ahead toward a very uncertain and possibly highly debilitated future, or backward to dwell on all I had lost, I pulled my focus in tight and concentrated simply on not panicking. It's a mind control trick that would later come to serve me well.

THE HUMAN BRAIN has been called the most complex structure in the universe. Since we know so little about the universe, there seems a reasonable chance this might be overstating the case, but there is no question that it's an impressive collection of cells. What we think of as reality is merely a construct of our brains. We can't begin to appreciate or comprehend to what extent our understanding is biased by our brain's filtering of all the data that streams in through our sensors.

If you think your senses provide a true interpretation of the world around you, then consider this: The conduction velocities of the electrical signals coming in from your various senses are not all the same. They don't arrive at the same time, and yet your brain adjusts things so you perceive them as doing so. If you watch your puppy nip your nose and then your toe, for example, you'd perceive, in both cases, that the sight of the nip and the feel of it occurred simultaneously, even though the travel time from the toe to the brain takes longer (about thirty extra milliseconds) than from the nose to the brain. And besides dealing with varying conduction distances, the brain must cope with different processing times. It takes almost five times longer for the brain to process sights than sounds (about fifty milliseconds as opposed to ten milliseconds). On the other hand, light travels through air about 880,000 times faster than sound, which is why you see the lightning strike before you hear it. Given these differences, it turns out that the sweet spot for humans, what's known as the *horizon of simultaneity*, is approximately thirty to fifty feet from the observer. At closer distances, what you hear precedes what you see, whereas farther away it's the reverse. Yet if you watch someone clap their hands, it doesn't matter if they're right in front of you or fifty feet away—in both cases, your brain will tell you that the sight and the sound occurred simultaneously. If you don't think this is utterly extraordinary, then you need to close your eyes and try to imagine your life playing out as a badly synced movie.

The point is that our brains aren't simply passive receivers of sensory input. How we experience the world is the result of a dialogue between the senses and the brain that integrates data from the external world with calculations and forecasts from our inner central processor. This dialogue evolved to enhance our chances of survival. We see, hear, smell, taste, and feel only what is needed. Much is hidden, but our ingenuity provides a capacity for revealing what is concealed from our immediate senses, if only we choose to *see*.

MY RECOVERY ROLLER-COASTER highs were few and far between, and when they did occur, it was difficult to trust them, knowing that they could be yanked away by some new complication at any moment. There was no single moment when I knew my vision would return. Rather, it was an achingly slow process. Initially, it was like seeing through a dirty, heavy lace curtain that moved around a lot, but if I stared long enough at something that didn't move, I could gather sufficient glimpses through my right eye to piece together an image. Eventually my sight came back enough that my father brought me a book to read—one he had selected for its length rather than its subject matter, so he was unaware that *Love Story* is about a young woman who dies of a blood disease. It was a short book, but it took me an inordinately long time to read; I had to hold my finger under each word, waiting for a passing hole in my blood curtain, but eventually I finished. And it felt like such a triumph to get through it that the downer ending had little impact.

My highest high came at the beginning of May, when I was finally allowed to sit up for the first time. To make this possible, I was fitted with a back brace that looked like a corset and was intended to provide support to compensate for my diminished muscle mass; I was told that the infection had eaten away 50 percent of the muscles in my lower back. The hole had begun to fill in with scar tissue, but there was still sterile packing that needed to be changed every couple of days. The day the doctor appeared in my room with the brace was one I had been anticipating for weeks. He slid the brace under me and cinched it up. Then, as I lay on my side with my legs extending over the edge of the bed, he rotated me into a sitting position. It was the first time I had been upright in three months. I was taking in all the sensations of being vertical when one of my favorite floor nurses walked by my room. I called out, "Hey, Adrienne, look at me!" and then I dissolved into tears, overwhelmed by happiness.

There was indeed much to be happy about; this outcome had been in serious doubt. As it turned out, my doctor's pronouncement that the spinal fusion had failed proved untrue. Instead, I was the beneficiary of a totally unexpected medical phenomenon: The massive infection in my back had produced increased calcification at the site of the fusion. Even though most of the bone chips had dispersed, a few had stayed put, and these had been enough to seed bone growth and create a solid fusion.

It was the end of May when I was finally allowed to leave the hospital. My sight had greatly improved, and the car ride was a visual cornucopia. I'd entered the hospital at the beginning of February; the trees had been bare and there was snow on the ground. Now everything was green and lush. The leaves on the trees vibrated with color and life. I felt desperate to drink them in. The heavy lace curtain in my right eye had broken apart and turned into floaters, dark spots that drifted through my field of vision—a distraction, but not a hindrance. My left eye was also progressing. A lot more light came through, although more around the edges than in the center. Turning onto our tree-lined street and seeing our two-story white clapboard house, with a magnificent spray of bright red tulips lining the walkway, I felt humbly grateful and elated all at once. Even the sight of the stairs leading to the front door and the longer flight to my bedroom on the second floor was exhilarating. Getting up those stairs the first time was tough, but with each day it got easier, and I was highly motivated. After four months of captivity, I was desperate to be outside. Five days after returning home, I managed to walk to my old climbing tree by the edge of the pond and give it a pat.

That tree is gone now, long ago chopped down and replaced by something more classically tree-like. My perception has outlived the reality. I can recall the shape and placement of its branches, the feel of its corrugated bark, the earthy smell around the base of the trunk where its roots extended toward the pond. Upon my return from the hospital, I saw that tree with much greater appreciation for the act of seeing, which made me think more deeply about *what* I was seeing.

We connect to nature through our senses, but that's not the only way to *see*. Over time I have added layers to the memory, greatly expanding my understanding of the how and why of treeness. Knowing how a tree carries water and nutrients from its roots to its leaves, and sugars from its leaves to its roots, and how willow bark harbors aspirin-like compounds adds layers of significance to the memory.

It is easy to understand why a tree is the living thing most often used to symbolize the human need to connect to nature. Poets and conservationists pen odes to "a tree called life" because it is a living being that nearly everyone has experienced firsthand. But how do we connect to nature on the much broader scale upon which we are now impacting it? We live on an ocean planet, but we have very little understanding of what that actually means. Ours is a living, breathing water world, filled with creatures whose existences are so utterly alien to our own, it's a significant struggle to relate to them.

How we choose to perceive the world around us shapes our existence. We believe we see the world as it is. We don't. We see the world as we need to see it to make our existence possible. At least that used to be true, but our world is changing so rapidly that we need a bigger picture of what makes life possible. We can't just look to trees to understand the intricate workings of the natural world. We must include the ocean and its myriad astonishments in a wider, wilder view. To look at the surface of the ocean without knowing the sparkling web of life that is woven through its depths is to be blind to its wonders and the part it plays in making our existence possible.

Chapter 2

FIAT LUX[*]

Spitting into my mask, I scrubbed the inside of the faceplate to prevent fogging. The sun had just disappeared below the horizon and the sky's light was rapidly fading. It was 2012, and I was off the Caribbean island of Saba with a bunch of folks who had come out on this dive boat because I suggested they might see bioluminescence, though I had never been diving here and couldn't be sure. We all donned snorkel gear and slipped from the boat into the warm tropical waters.

Floating at the surface, I eagerly scanned the sandy bottom ten feet below. Out of the corner of my eye I thought I saw something. Was that a glowing spot in the sand? When I looked directly at it, it seemed to disappear. But a little while later there was another, and another. Then I heard somebody shout, "Hey—look at that!" Someone else managed a muffled "Wow!" into their snorkel as more and more of the blue lights appeared and then began rising off the seafloor like strings of glowing champagne bubbles. In just moments we were surrounded by the ephemeral chasing lights that are the mating displays of sea fireflies.

These remarkable little creatures are crustaceans called ostra-

---

[*] Latin phrase meaning "Let there be light" (the more literal translation is "Let light be made"), as seen in Genesis 1:3.

cods, which aren't much bigger than sesame seeds but are capable of producing copious amounts of light. Their common name comes from the fact that they use their bioluminescent displays as fireflies do on land—to attract mates. It's the males that put on the show; they emerge from Caribbean reefs, seagrass meadows, and sand just after twilight, and as they swim through the water they squirt out discrete globs of light that are a mix of their light-producing chemicals held together with a dab of mucus. The lights appear and then disappear sequentially. For some of the people on that boat, it was the first time they had ever seen bioluminescence, and they were totally captivated and awash with questions.

The first time I saw bioluminescence was in my backyard. Mine is a quintessential childhood memory: fireflies flashing their lovers' code on a warm summer's eve as I ran barefoot in pursuit, stirring up the heady green smell of freshly mown grass. Catching these living lanterns was easy. I could hold one in my cupped hands and peek between my palms to watch its abdomen flare. *How does it do that?* I marveled. It's a question not easily answered. If you try to take a firefly apart to see how it works, the first thing you have on your hands is a nonworking firefly.* However, holding it gives at least one clue to its sorcery. Bioluminescence is cold light. This seems surprising, because, based on our everyday experiences with the sun, candle flames, and incandescent lightbulbs, we associate light with heat. Nevertheless, they are not inseparable.

All light comes from atoms. Picture the simple Bohr model of an atom, where negatively charged electrons orbit a positively charged nucleus and the orbits are represented as concentric shells. Different distances from the nucleus represent different energy levels. Electrons in the orbits nearest the nucleus have the least energy. If an electron absorbs enough energy, it jumps to an outer orbit and

---

*Douglas Adams made this astute observation about cats, but it applies equally well to fireflies.

then, when it falls back down to its ground state, it gives up energy as a packet of light called a photon. All light is generated by this same basic process. The only distinction between different kinds of light is how the electrons get excited in the first place.

In a candle flame, say, or an old-school lightbulb, the electrons are excited by thermal energy, in which case the light is called incandescence. The prevalence of such sources is why we associate heat with light. But there are other means of excitation, such as a chemical reaction, in which case the light is called chemiluminescence. Bioluminescence is a special case of chemiluminescence, distinguished by the fact that the light-producing chemicals are synthesized by living organisms. Light sticks, which emit light but not heat, are another example of chemiluminescence, one where the chemicals are manufactured by humans.

Any light that isn't caused by incandescence is lumped under the heading of luminescence. Besides bioluminescence and chemiluminescence, there are more obscure phenomena like sonoluminescence, caused by sound, and triboluminescence, caused by the breaking of chemical bonds.* Two more common examples of luminescence, fluorescence and phosphorescence, frequently get confused with bioluminescence, but they're not equivalent, because their excitation energy comes from light rather than from a chemical reaction.

Something that is fluorescent absorbs light of one color and re-emits a different, longer-wavelength (i.e., lower-energy) color. Black-light posters, for example, absorb ultraviolet light that is largely invisible to our eyes—which is why it's called black light—and re-emit it as a visible color. A fluorescent lightbulb is so called because

---

* As a demonstration of triboluminescence, get yourself some Wint-O-Green Life Savers (make sure they aren't sugar-free) and then convince a friend to join you in a dark room so they can watch you chew the Life Savers with your mouth open and tell you what they see. Or you can forgo this bonding experience and just crush the mint with a pair of pliers.

the inside of the glass tube is coated with a fluorescent material that absorbs the UV photons emitted by gas atoms in the tube, and emits visible photons. This is accomplished with almost no release of infrared light (heat), which is why you can touch a fluorescent bulb without being burned.

Phosphorescence is *not* bioluminescence, despite how often you hear the words equated. It's a very common misconception that has been repeated so many times it borders on a disinformation campaign. Phosphorescence, like fluorescence, is excited by light, but with an added delay in the re-emission, making it the basis of the fearsome glow-in-the-dark decorations and toys sold at Halloween. Part of the confusion between phosphorescence and bioluminescence stems from descriptions of bioluminescence as resembling "liquid phosphorus." Certain forms of chemical phosphorus produce a dim glow, and as a result, the word *phosphorescence* was originally coined to describe things that glow without burning, but, in fact, the glow of phosphorus is *not* phosphorescence but rather the result of a chemiluminescent reaction.

I've often thought that it's too bad there isn't an alternate word for bioluminescence, because the fact that people find it difficult to spell and pronounce has contributed to its relative obscurity. Years ago, I worked with an artist to create a coloring book about bioluminescence that included glow-in-the-dark paints. Since the whole idea was to share the feeling of pure wonder, I struggled to find just the right title. I thought *The Living Lights Coloring Book* would do but then rejected it because it might be mistaken for something to do with religion and went with a more scientifically accurate but less approachable title, *The Bioluminescence Coloring Book*—which may be why we still have several thousand of these masterworks in storage.

The names of the chemicals that produce the light in a bioluminescent reaction are a bit more manageable: luciferin and luciferase. That terminology was the invention of the French physiologist Raphaël Dubois (1849–1929), who is generally credited with ushering in the modern era of bioluminescence research. Working with a bioluminescent click beetle and later a bioluminescent clam, he demonstrated that their light-producing chemicals could be extracted using an experimental approach involving hot-water and cold-water extractions. When the tissues were ground with cold water, he got light emission for several minutes before it went dark. Hot-water extracts, on the other hand, did not produce light, but when the now-dark cold-water extract was mixed with the hot-water extract, he discovered that he could reactivate the light emission.

Dubois called the substance from the hot-water extract luciferin and the substance from the cold-water extract luciferase. The terms were derived from *lucifer,* which means "light-bringing" in Latin (from *lux,* or "light," and *ferre,* "to bring"). The suffix *-ase* is traditionally used to name enzymes. Enzymes are large, complex molecules that are destabilized by heat, while substrates are generally much smaller, more stable molecules. Therefore, Dubois made the following deductions: (1) While both the enzyme and the substrate were initially present in the ground-up cold-water extract, the substrate, which he called luciferin, was used up in a matter of minutes and the light extinguished. (2) The hot-water extract did not produce light initially, because the heat denatured the enzyme, leaving just the heat-stable luciferin. (3) Mixing the hot- and cold-water extracts was therefore the equivalent of mixing the heat-stable substrate, luciferin, with the heat-labile enzyme, luciferase.

Dubois's terminology is still used today, but it sometimes causes confusion because people believe the names refer to specific chemicals. They don't. They are generic terms used for any bioluminescent substrate or enzyme, of which there is a surprising variety.

That there are so many different bioluminescent chemicals is a testament to how important bioluminescence is. The ability to produce light is so critical to survival that it has been selected for independently more than fifty times during evolutionary history. This is known as "convergent evolution," where creatures that aren't closely related evolve similar traits in order to adapt to similar circumstances. For example, although sharks and dolphins share analogous streamlined body shapes and fins of comparable form and function, this is not because they are closely related genetically; sharks, after all, are fish, while dolphins are mammals. Rather, that particular body plan works well for maneuvering through water and therefore provides an advantage—allowing them to catch more food and evade more predators and thus survive long enough to pass on their DNA.

In the case of bioluminescence, many very different animals solved the problem of how to survive in the dark the same way: *Make your own light.* In textbooks on evolution, the classic example given for convergent evolution is eyes, like those of squid and octopods (invertebrates) and those of fish and humans (vertebrates). In both cases, the eyes are camera-like in that they have an iris and a lens at the front that focuses light on the photoreceptors at the back. However, while the photoreceptors in the cephalopod eye face toward the lens, for vertebrates they face away—clear evidence of their independent origins.

In fact, eyes have evolved independently more than fifty times, appearing in diverse animals like jellyfish, flatworms, flies, mollusks, fish, and whales, in a variety of forms, from simple pits or eye spots to more elaborate camera-like eyes and complex compound eyes (sometimes called bug eyes) consisting of thousands of individual light-gathering units. This is similar to how many different times it's thought that bioluminescence evolved. However, there is one remarkable distinction. All eyes depend on the same chemistry— a light-sensitive protein called an opsin—while the luciferins and luciferases that make bioluminescence possible are unique in different groups of animals.

The independent origins of such disparate chemical systems in so many different groups of animals are not only a stunning testament to just how critical bioluminescence must be to their survival but a treasure trove for science. Like Prometheus stealing fire from Zeus to give to mankind, scientists have found all manner of ways to harness living light—using chemicals extracted from bioluminescent organisms to illuminate the inner workings of cells and test for life processes and key molecules.

One chemical, green fluorescent protein (GFP), extracted from a bioluminescent jellyfish, has so advanced human understanding of cell biology that the impact of its discovery has been equated to the invention of the microscope. Bioluminescent sea fireflies have provided the means to image tumor tissues and test the effectiveness of anticancer agents in a single animal, instead of having to sacrifice large numbers of animals at different times to study the effectiveness of the treatment. The bioluminescent chemistry of terrestrial fireflies is routinely used to test for bacterial contamination and, less routinely, to test for the presence of life on Mars. There are many more examples. There are also remarkable numbers of bioluminescent chemistries still awaiting discovery, and new applications to be invented that could lead to equally phenomenal breakthroughs.

However, even where we have identified what chemicals are involved in producing bioluminescence, that still doesn't answer the question *How does it do that?* Saying that we understand *x* because we know *y* would be akin to claiming that you understand how a car works because you know it runs on gasoline. There's a bit more to it than that.

The fact that I became involved in research that asked that very question of a different light-producing organism had nothing to do with my youthful musings on the subject. In fact, I think it's safe to say that I gave no thought whatsoever to how animals make light for almost two decades after my childhood firefly encounters. Rather, my obsession with bioluminescence grew out of my brush with blindness.

*   *   *

I SAW THE world very differently after I got out of the hospital. There were moments of surpassing joy for the act of seeing, but there were also moments of unfamiliar doubt. I had regained sight but lost the supreme confidence of youth that *anything* is possible. I learned the hard way that the two sides of the *anything* coin might include *bad* just as easily as *good,* which meant I felt the need to consider possible negative outcomes and try to always have a plan B.

It is a measure of just how much my worldview had shifted that when I returned to Tufts in the fall of my sophomore year, I changed my major from marine biology to premed. As it turned out, this was a temporary detour, but at the time it was a significant upheaval of my life's plan. I had held firm to the goal of becoming a marine biologist since I was eleven years old.

That year, when kids my age would normally be in sixth grade, was transformative for me. Up until then I had been a mediocre student. I had always hated school and, as a result, had paid little attention to anything my teachers were saying, merely biding my time by daydreaming until I could get home and be outside. But the year I turned eleven was a year of travel that woke me from my reverie. My parents were Ph.D. mathematicians, and that year was a sabbatical year for my dad, who was a professor at Harvard. My mother, who had given up full-time teaching to raise my brother and me, resigned her part-time position at Tufts to spend the year abroad.*

My brother, eleven years older, was married by the time I was ten, so it was just me and my parents that year. The plan was to spend half the year traveling and half in Australia, where my dad had a Fulbright fellowship at the University of Melbourne. Since I would be out of school for most of the year, Mom and Dad would be my teachers. Homeschooling wasn't a thing in those days, and this was considered a bit unorthodox. However, since the primary focus of

---

* Several years after our return, she resumed teaching full-time at the University of Massachusetts Boston.

the sixth-grade curriculum was world history and math, and I would be traveling around the world visiting historical wonders with a couple of mathematicians, my school grudgingly conceded that I might survive the academic deprivation and ultimately decided that I would be allowed to come back into the public school system without having to stay back a year.

The things I saw during our travels awakened me to a whole world of possibilities, and my childhood daydreams of being a female Zorro transitioned to more-adult ambitions. Magnificent art in Europe, archaeological wonders in Egypt, wrenching human suffering in India, and fantastic wildlife in Australia sequentially inspired in me new career goals of artist, archaeologist, humanitarian, and biologist. Australia had the greatest impact, reinforcing a lifetime fascination with animals through encounters with koalas, kangaroos, wallabies, wombats, fairy penguins, black swans, emus, flocks of brilliantly colored parrots, and that most absurd of creatures, a platypus.* It seemed impossible to imagine anything more fantastic than the chimeric absurdity of a duck's bill, a beaver's tail, and webbed feet like an otter's, until I learned that the female lays eggs like a reptile and feeds her babies on milk like a mammal, while the male exhibits his studliness with venomous spurs on his hind legs.

In Fiji, where we stayed on the aptly named Coral Coast—our last stop before heading home—my dream of becoming a *biologist* morphed into *marine biologist*. Our thatch-covered hut, which sat right by the water, had open windows and a canopy of mosquito netting over each bed. During the day, I was allowed to roam the coral reef on my own. Now I look back with dismay at our ignorance and that of the resort's owners, who not only permitted but encouraged tourists to don sneakers and walk out on the flat-topped live coral reefs at low tide. At that time, the reef was still magnificent, but I

---

*Only one because nobody can agree on what the plural of "platypus" should be.

have no desire to return these many years later, because I know what I would see would be only a faint shadow of its former glory.

The reef was a rainbow-colored kaleidoscope of life so rich in natural wonders that I could never focus on one thing for any length of time before my attention was yanked away by some other fantastic sight. There were pink, purple, and gold corals forming interleaved plates and reticulated domes. There were deep, glass-clear tide pools, each like a separate tropical aquarium filled with brilliantly colored fish, fantastic banded shrimp with elegant long antennae, cobalt-blue starfish, and giant clams so big they could have swallowed me whole, each sporting an exquisite scalloped mantle colored brightly in azures, greens, indigos, and golds that seemed to glow from within.*

In one shallow pool, I found an especially exotic-looking fish with burnt umber and white stripes and fin rays sticking out from it in all directions. Unlike the other tide-pool fish that darted away when I tried to peer closely at them, this pugnacious fellow just fluffed out his fins and stared up at me as if to say, "Yeah? So what's your problem?" My *problem* was that I desperately wanted to share this fantastic find with my parents, but I was a long way from our hut. I was afraid I wouldn't be able to find him again if I tried to lead them back, so I very carefully herded him into a plastic bag I had with me and started to carry him. Then I got worried that he might suffocate and, knowing how terrible that would make me feel, I gently put him back.

Years later, I would see an identical fish at an aquarium, with a sign over his tank explaining that this was a lionfish, who sported elaborate coloration intended to advertise his venomous spiky fin rays. Had I been less gentle with him, my career as a marine biologist might have ended before it began. As it was, passion for observ-

*This glow is sometimes mistaken for bioluminescence but is in fact sunlight-induced fluorescence.

ing living oddities like this strange fellow motivated me to become a
better student and get the grades I needed to become a marine biologist.

To give up that long-held dream after my hospital experiences
should have been wrenching, but it wasn't—at least not at first. With
my life-and-death struggles still raw, I felt that I needed a greater
sense of security, and while nobody could tell me exactly what I
needed to do to become a marine biologist or what kind of career or
job prospects I could expect, the "becoming a doctor" track was well
worn and clearly marked.

My favorite premed class was human physiology. Physiology is all
about how living organisms work, which fascinated me. I liked it so
much that I signed up for another course as an elective, "The Physiology of Behavior." Ned Hodgson, who taught this class, was an excellent teacher and consummate storyteller. He would interweave
lectures on the neural basis of behavior with personal anecdotes that
were usually funny and sometimes inspirational. In one class, he
shared his experience of making a breakthrough about how insects
detect chemicals in their environment—a finding that was published in the prestigious journal *Science*. I sat on the edge of my chair
as he described what it felt like to discover something that no one in
the history of the world had ever known before. The sense of wonder
and exhilaration that he conveyed was palpable, and I knew I wanted
to experience that thrill.

That class reignited my fascination with understanding animals.
As a result, I said goodbye to the premed track and, during my junior
year, enrolled in another course that Ned had helped develop: "Tropical Marine Biology." It was taught during Tufts's January mini-
semester at Lerner Marine Laboratory, located on the Bimini Islands
in the Bahamas. A field station of the Smithsonian, this place had
everything an aspiring marine biologist could want: gin-clear tropical waters full of all manner of marine life—neon-colored tropical
fish, moray eels, barracuda, and lemon sharks, as well as our own

resident dolphin, named Charlie Brown. It was a full-immersion (pun intended) course, one of the first of its kind. There were some classroom lectures and some field excursions led by Ned and our other instructors, but mostly we learned about the coral reefs, the mangrove habitat, and the seagrass meadows by spending hours and hours swimming through them.

After months of confinement in bed and then many more months of limited mobility in a back brace, I found the physical freedom of floating weightless underwater intoxicating. I was healthy again. My vision was back. The leg pain was gone. My back ached sometimes but was one hundred times better than at any time I could remember. I recall one dive during which this new reality hit me with such a rush that I felt like Scrooge after the last of the spirits leaves and he realizes he's still alive. "I must stand on my head! I must stand on my head!" he shouts as he flings his legs in the air—a feat easily accomplished underwater, except that I was so giddy with the pure joy of it that I started laughing, causing my mask to fill with water. I cleared it and proceeded with the dive, only slightly chastened and still joyful.*

When the course drew to a close and it came time to leave, we were all devastated. Going back to New England in the dead of winter was the worst kind of tropical decompression. Also, there was this other little matter that had been preying on my mind: Ten days after my return, in the middle of my junior year of college, I was to be married.

DAVID AND I began dating at the end of our senior year in high school. He was smart and funny and was on the gymnastics team. Over the course of that summer, I taught him to water ski, and he

---

*I wasn't "narced," i.e., suffering from nitrogen narcosis, which can result in such euphoria that afflicted divers have been known to offer their regulator to a passing fish that looks in need of oxygen. I was only at twenty feet and have made much deeper dives without experiencing that syndrome.

taught me how to kiss. But then he joined the Navy and I started at
Tufts. We were an unlikely match. I came from a small upper-
middle-class academic family. He grew up in a large blue-collar family: his father a disabled firefighter; his mother also disabled, by
polio; plus five rowdy kids in a three-bedroom, one-bath apartment.
Nevertheless, we stayed in touch through letters. Immediately after
my surgery, he ran up an enormous phone bill, which he could ill
afford, talking to either my mother or his mother daily, trying to get
updates on my status. He also wrote to me *every* day. I had to have
these letters read to me, which was sometimes awkward, because
unlike most men, David had no difficulty expressing the depths of
his emotion, either verbally or in writing. That skill has since helped
contribute to our long and happy marriage, but at the time I found it
overwhelming and acutely embarrassing.

After boot camp and medical corps school, David was assigned to
Chelsea Naval Hospital in Boston and we continued seeing each
other. It was thanks to him that I was able to start scuba diving again
a year after I got out of the hospital. Without a boat to use, we would
drive up to Gloucester or down to Plymouth to dive off a rocky shoreline. Postsurgery, I wasn't supposed to carry anything heavy, so
David had to lug all of our gear over the rocks and down to the shore,
and then he had to help put the tank on my back once I was in the
water.

David proposed just before I started my junior year at Tufts. At
least I *think* he did. We were on a camping trip and bedded down by
the campfire; David was waxing poetic about what a great trip this
was and how we should have more like it and *wouldn't it be great if we
could wake up together every morning?* I assumed he was being rhe-
torical and apparently drifted off to sleep before he got to the point.
Never lacking in confidence, he took my silence as assent. I found
out I had been proposed to when we returned home and I heard him
tell his mom that we were engaged.

Getting married that young wasn't in my game plan. I wanted to get married *someday*, but I intended to wait until after I got my Ph.D.—as my parents had. Also, David was the first boy I ever kissed! There was no question that I was crazy in love with him, but how could I really be sure he was *the one* for me? Nobody gets that lucky on their first go, right? On the other hand, what if he *was* the one and I blew it because I couldn't believe my own good fortune? If the hospital had taught me anything, it is that life is capricious, and you need to appreciate what you've got while you've got it.

These were the thoughts drifting through the back of my brain as I was growing gills in Bimini. Meanwhile, back at home, David was making all the arrangements for our wedding with the help of our mothers. It was to be a small chapel ceremony with just immediate family and a few close friends. I was so ambivalent about the whole thing that I hadn't even shopped for a wedding gown before I left for Bimini; I had just arranged to borrow an ill-fitting blue bridesmaid dress from my future sister-in-law.

David had planned to meet me at the airport, but when I got off the plane he was nowhere in sight. As I walked toward baggage claim, I felt all those questions and doubts about marriage moving from the back of my brain to front and center. Then I spotted him— bounding up the escalator three steps at a time, looking incredibly handsome in his Navy peacoat. He picked me up and spun me around so hard that the centrifugal force wrenched all those questions and seeds of doubt right out of my head, with the result that, although I never technically said "yes," I did say "I do." And it was the smartest thing I've ever said.

AFTER I GRADUATED from Tufts and David got out of the Navy, in 1973, my path to bioluminescence followed a zigzag track. We both got jobs in the Boston area—he at W. R. Grace chemical company and me at Harvard Medical School. We spent two years working as lab techs, then loaded up our car and headed for Santa Barbara, where David had been accepted into a bachelor's program at Brooks Institute of Photography and I would be starting a master's program in electrical engineering at the University of California, Santa Barbara.

My interest in electrical engineering was grounded in my belief in the importance of having a plan B. Plan A had solidified into getting a Ph.D. in neurobiology, with the idea that I could study the physiology of behavior of marine organisms. Plan B grew out of my increasing interest in electronics, which I had been studying a little on my own. I figured that if I went with plan A, then the background in instrumentation would undoubtedly prove valuable in neurobiology; and if for some reason I decided not to go on for a Ph.D., then a master's degree in instrumentation would probably make me more employable than one in biology, and maybe I could even work in ocean instrumentation.

Several weeks after starting at UCSB, though, I realized that I needed to come up with plan C. The degree program seemed different from the multidisciplinary one I had read about. My advisers were hell-bent on channeling me into systems engineering, which they believed was a much better fit for my stated goal of working in neurobiology. These were frustrating conversations; I constantly felt that I was being preached at rather than listened to. I was a couple of weeks in before I discovered that the program I had read about was actually offered through the physics department. I went immediately to Virgil Elings, the founder and director of the scientific instrumentation program, and asked to transfer. I had already determined that there were still openings available, so I was taken aback when he told me he didn't believe women belonged in instrumentation, because, as he put it, "women don't know how to tinker."

It wasn't the first time I'd had to deal with this kind of thing. My former boss at Harvard Medical School liked to hold forth on his belief that women lacked the spark of genius required for true innovation and that their proper function in life was to act as stabilizing agents for the heady brilliance of men like himself. This is standard pushback for anyone trying to break through social barriers. It is doubly devastating if you have to fight against your own socially embedded doubts at the same time that you need to be boldly proclaiming your ability to accomplish the task at hand. I was extraordinarily fortunate that I had a secret weapon: my mother as a role model. She was a woman who grew up on a farm in western Canada, could handle a team of four horses pulling a plow, and went on to get a Ph.D. in mathematics from Bryn Mawr College.* Although I sometimes had doubts about my own abilities, my mother taught me, by example, to never waste time thinking my inadequacies had anything to do with my sex.

I could have countered Virgil's argument with the fact that I *love* to tinker; I used to tinker with our old outboard motor, and David and I had rebuilt a couple of Volkswagen engines together while we were dating. But with the exhausting arguments of recent weeks fresh in my mind, I didn't feel up to carrying the women's rights banner over these particular bastions. You need to pick your battles in life, and since this one related to plan B, not plan A, I decided to regroup and redirect. I transferred to biochemistry.

When I completed my master's degree two years later, David still had a year left before he would finish at Brooks. I was planning to apply to graduate schools back east for my Ph.D., but in the interim, I needed employment. I decided to approach Jim Case, a neurobiologist whose graduate course in neurobiology I loved and had aced.

JIM CASE'S TWO most notable attributes were his round, bald head—he looked sort of like a melon wearing glasses—and his dry

---

* One of the stories I grew up on was about how, after receiving her B.A. in mathematics and a University Gold Medal for high honors in mathematics, my mother went home to help out on the farm. One day she was in the fields, working a team pulling a binder, when the harness broke. The neighbor in the next field saw she was in trouble and rushed over to help, but by the time he got there she had gotten control of the team, repaired the harness with a piece of hay wire, and was back on the binder. He looked at her and said, "Well, I guess it's all right for a girl to study mathematics as long as you can still do something useful."

wit. Jim's bland, bespectacled baldness, coupled with his standard attire of a sweater vest and tie, belied his rapier wit, sometimes with wicked effect. At first, his response to my job inquiry seemed to be one of his jokes. He said, "Graduate students are cheaper to pay than research assistants." It took a minute to sink in that he was offering me a paid graduate-student position in his lab. He explained that he and Beatrice Sweeney, who worked in the lab one floor up, had been discussing the possibility of finding a prospective Ph.D. student to study the electrophysiology of a bioluminescent dinoflagellate that Sweeney had isolated and was maintaining in culture. He then proceeded to send me upstairs to talk with her about the project.

With her full head of beautiful cropped white hair and her standard casual attire, which included year-round flip-flops and a toe ring, Beatrice Sweeney was the antithesis of Jim Case. Known to her students as Beazy (while Case was always referred to as Dr. Case), she was energetic and intense, and she focused that intensity on me as we sat in her office and she talked enthusiastically about something called "bioluminescence," while I attempted to camouflage my cluelessness by keeping silent and nodding at what I hoped were the right moments.

I had a pretty vague understanding of what the word meant, but I didn't feel I needed to let her know that. After she had talked for a bit, Beazy led me into her lab, where she opened the door of an incubator that looked like a tall refrigerator and pulled out a large Erlenmeyer flask with a cotton plug in the top and a couple of inches of liquid in the bottom. She explained that this was a culture of a dinoflagellate called *Pyrocystis fusiformis,* a beautifully descriptive name that means fire (*pyro*) cell (*cystis*), shaped like a football (*fusiformis*). She held the flask up to the light and pointed out that these single cells were so large, you could see them without a microscope— barely. This was significant because the idea behind this particular project was to stick an electrode inside one of these cells and record the electrical activity that triggered its bioluminescent flash. She then proceeded to turn the lights out and swirl the flask, to spectacular effect. Dazzling blue light flashed forth in a whirlpool of liquid brilliance that lapped around the edges of the flask, illuminating her face and causing me to gasp.

The most natural question in the world when seeing something like that is, once again, *How does it do that?* And that was the question my graduate research was supposed to answer! I was hooked.

FIRST FLASH

When I got home that evening, after sharing with David the surprising result of my job interview, I looked up *bioluminescence* in our recently purchased *Encyclopedia Americana*. It was a short entry—less than a half page of text—that defined it simply as "chemical light produced by living organisms." There was very little explanation of how the light was produced, except that it could manifest as a steady glow or in flashes, and that it involved a substrate and an enzyme and it was thought that these chemicals were different in different species. Living light producers on land included the well-known fireflies and less well known worms and fungi, the latter illustrated by a picture of green-glowing bell-shaped toadstools. Also mentioned, but not shown, were deep-sea fish with glowing lures used to attract prey, as well as squid and crustaceans that can spew luminous secretions, just like an octopus releasing an ink cloud as a defense against predators. Additionally, there were microscopic light emitters that included bacteria as well as dinoflagellates like those I saw in Beazy's flask. At the bottom of that encyclopedia entry, I was impressed to see that its author was Beatrice M. Sweeney.

As I would later learn, bioluminescence was a field in which both Beatrice Sweeney and Jim Case were superstars. Beazy had done groundbreaking research on circadian rhythms using bioluminescent dinoflagellates to learn how temperature and light affected their internal clocks, while Case had focused on the neurophysiological control systems of fireflies. Their idea that *P. fusiformis** might make a good model organism for studying bioluminescence had come out of recent pioneering work by a scientist named Roger Eckert, who had managed to stick an electrode in a very different bioluminescent dinoflagellate, one with an equally wonderful name: *Noctiluca scintillans*, which means "sparkling night light." While the notion of getting a microelectrode inside a dinoflagellate sounded challenging, I had the reassurance of knowing it was possible, at least for *N. scintillans*, the largest of all dinoflagellates.

JUST BECAUSE DINOFLAGELLATES are single-celled organisms doesn't mean they're simple. In fact, their diversity and peculiarities can be a bit overwhelming. "Dinos" occupy a range of habitats, including marine, freshwater, and estuarine waters, although they are predominantly (about 85 percent) marine. Some species live on snow and sea ice, while others are parasitic on animals like crustaceans and fish, and still others grow as endosymbionts in corals, bestowing upon their hosts both life-sustaining energy and a resplendent palette of colors. For reasons unknown, they contain more DNA on a per-cell basis than humans do, in some cases almost one hundred times more.

Some dinoflagellates produce toxins and can occur in such profusion that they turn the water red, earning the label "red tide," and if their toxins accumulate in shellfish or fish that are eaten by humans, the result can be potentially deadly outbreaks of paralytic shellfish poisoning (PSP), neurotoxic shellfish poisoning (NSP), diarrhetic

---

* Scientists, as a rule, are pretty picky about the naming of things. Every organism is given two names. The first is the genus and is always capitalized; the second is the species and is never capitalized. In scientific publications (and in this book) both names are italicized and written out the first time they appear. Subsequently, either to save ink or to cut down on writer's cramp, the genus is abbreviated to a single letter.

shellfish poisoning (DSP), and ciguatera. In fact, lowly dinoflagellates are responsible for ten times more human fatalities each year
than shark attacks.*

Dinoflagellates are so highly varied that it sometimes stretches the
imagination that they are related, ranging in size from twenty to two
thousand microns and in shape from smooth, round, and unarmored
to spiked, armored,† and sporting two beating flagella—lashlike appendages used for propulsion. It is the latter form that is most typical
and from which the dinoflagellates' name derives, after the Greek
dinos, which means "whirling," and the Latin *flagellum,* "small whip."
It is this derivation that spurred purists like Beazy to insist the correct
pronunciation is *DEE-no,* not *DYE-no,* to distinguish it from a word
like *dinosaur,* which is derived from the Greek *deinos,* meaning "terrible," and *sauros,* "lizard." Dr. Case, unmoved by Beazy's classicism,
insisted on *DYE-no,* forcing me to switch pronunciations depending
on my audience or, when they were both in the room, to either avoid
the word or mumble through it.

That some dinoflagellates produce light adds to their mystique;
the light-producing dinoflagellates are known as sea sparkle. Blooms
are responsible for breathtaking light shows as flashes are elicited by
the slightest touch or disturbance in the water, creating eddies of
molten light and cold blue fire in every wave cap. Charles Darwin
provided a vivid description of the appearance of one such bloom
witnessed from the deck of the HMS *Beagle* as it cruised off the coast
of Uruguay:

> While sailing a little south of the Plata on one very dark night,
> the sea presented a wonderful and most beautiful spectacle.

---

*To be fair to sharks (and dinoflagellates, for that matter), over the past decade
the worldwide average of fatalities from shark attacks has been six per year.

†Armor in dinoflagellates consists of cellulose plates that form a protective shell.
Some plates are smooth, but some sport elaborate pores and grooves that make
them look very much like the armor plating on a crocodile.

There was a fresh breeze, and every part of the surface, which during the day is seen as foam, now glowed with a pale light. The vessel drove before her bows two billows of liquid phosphorus, and in her wake she was followed by a milky train. As far as the eye reached, the crest of every wave was bright, and the sky above the horizon, from the reflected glare of these livid flames, was not so utterly obscure as over the vault of the heavens.

Such phenomena are more common than most people realize. Unfortunately, artificial lighting used on boats and human habitation along shorelines, where dinoflagellate blooms are most likely to occur, overwhelm bioluminescence. As a result, modern sailors have less poetic encounters with living light, often stumbling upon it for the first time in the ship's head (a.k.a. toilet), which is flushed with unfiltered seawater. The result is that many a seasick sailor who was so toilet-huggingly sick that they neglected to turn on the lights may have thought they were having a religious experience while "talking on the porcelain telephone to God."*

Blooms, which can reach concentrations of millions of cells in a cup of water, are largely unpredictable but often appear after the introduction of nutrients, as may occur with rain runoff. There are a few very special places called bioluminescent bays, where bioluminescent dinoflagellates are present in high abundance year-round. It takes a unique combination of characteristics to sustain such densities. Requirements include a tropical climate and a shallow bay with a narrow channel and small tidal flux; a dense stand of healthy mangroves around the edges of the bay; and prevailing winds that work to increase the residence time of the dinoflagellates in the bay. These magical places, not surprisingly, attract a lot of tourists, often to the

---

*One of a seemingly endless array of euphemisms for seasickness, which are amusing to everyone but the victim.

detriment of the bays. Many have been negatively impacted by a variety of anthropogenic stressors like light pollution, sunscreen chemicals, motorboats,* coastal development, and pollution runoff from construction sites, roads, and parking lots.

Dinoflagellate bioluminescence was well known to early mariners, though its cause was not. Aristotle (384–322 B.C.) described the light, which appears when the ocean is agitated at night, as akin to lightning. Some two thousand years later, Benjamin Franklin (1706–1790) drew a similar analogy. In fact, he believed the sea was the *source* of lightning. Based on his observations of sparkling seas, he assumed that the light was a kind of electric fire resulting from friction between water and salt. However, excellent scientist that he was, he began to question this view when he conducted experiments demonstrating that a sample of sparkling seawater in a bottle would produce light when first shaken but would lose the capacity over time. He also found that if he added sea salt to freshwater, he could not produce any light. Based on these results, he said, "I first began to doubt of my former hypothesis, and to suspect that the luminous appearance in sea water must be owing to some other principles." This new perspective was reinforced by a letter Franklin received from the then governor of Massachusetts, James Bowdoin (1726–1790). A keen observer of nature himself, Bowdoin described how he discovered that the sparkles in seawater could be removed by filtering the water through a cloth, and he suggested to Franklin that "said appearance might be caused by a great number of little animals, floating on the surface of the sea." Franklin concurred and accordingly discarded his original hypothesis that lightning came from the sea.

---

*Motorboats stir up sediments that can increase algae growth due to nutrient resuspension and decrease sunlight for dinoflagellates. A two-stroke engine, like the one I grew up water skiing behind, can release 25 to 35 percent of its unburned gas and oil into the water. Which explains why, when motorboats were eventually banned on the pond I grew up on, it became a much healthier ecosystem with cleaner water and abundant birdlife.

Dinoflagellates aren't animals, but they aren't plants, either; they are part of a large grouping of mostly single-celled life-forms called protists (eukaryotes that are not a true animal, plant, or fungus; examples include amoebae, paramecia, and algae). About half of all known dinoflagellates behave like plants and get their energy from photosynthesis, while the other half behave like animals and get their energy from consuming other organisms. *N. scintillans* is one of the animal-like dinoflagellates, while *P. fusi** is plant-like. Part of the point of my thesis research was to determine how different or similar they might turn out to be in terms of their light-producing abilities.

This is the kind of research that often mystifies nonscientists. Why does it matter? Answering that question comes down to the difference between basic science and applied science. With applied science, there is a specific problem that needs to be solved, like how to prevent polio, how to cure cancer, or how to build a bigger bomb. With basic science, there is a curiosity-driven question to be answered, like *How does a living creature make light?* In the latter case, there is no specific application in mind. It is simply driven by the fundamental human desire to understand how things work. Some of the greatest scientific discoveries ever made were generated by basic science, and all applied science is built on the foundation of basic science.

When I began my thesis research, the luciferin and luciferase involved in light production were as yet undescribed in dinoflagellates. All that was known was that the light originated from organelles (membrane-bound structures inside cells) called scintillons. When you bump a dinoflagellate, its scintillons flash. The question was how?

Every living creature has some means of responding to environmental changes. The organ or cell or system that effects (i.e., brings

---

* A further abbreviation—unauthorized but commonly used.

about) that change is called the effector system. When you absent-mindedly lift the lid on the casserole dish that minutes ago you re-moved from a four-hundred-degree oven, but this time without benefit of hot mitts, a remarkable sequence of events occurs before the cursing ensues. You actually pull your hand away *before* your brain registers the pain. To accomplish this, a sensory neuron de-tects a potentially damaging stimulus and transmits a nerve impulse to the spinal cord, where a relay neuron transfers the signal to a motor neuron. The motor neuron then sends a nerve impulse back out to the effector, in this case the muscle, causing it to contract and pull your hand away. At the molecular level, the muscle cells con-tract because two large molecules called actin and myosin ratchet past each other. As in most effector systems, these large molecules are poised to act when triggered by the introduction of very small charged atoms, called ions.

The flash of a bioluminescent dinoflagellate is another sort of ef-fector system, in which a mechanical stimulus, like bumping the cell, initiates an electrical signal that somehow triggers light emis-sion from the scintillons. Just like the withdrawal reflex that protects us from dim-witted interactions with hot casseroles, it's a response that occurs without any conscious effort, but in this case it all occurs in a single cell.

Life throbs, pulsates, and scintillates because of excitable mem-branes. We owe our mobility, our thought, our very existence to the ability of a cell to transmit an electrical signal, but this is a very dif-ferent electrical transmission than occurs in electronics, which are so named because they transmit electrons. The electrical signal in cells is the result of a flow of ions across membranes.

In a classic neuron, which sports a long, slender projection called an axon, the opening and closing of sodium and then potassium ion gates ripple down the length of the axon in less than the blink of an eye. Although that sounds fast, it's actually glacially slow compared with the speed of an electrical signal, but in evolutionary terms, all that's required is sufficient speed to evade one's predators. Take, for example, the giant nerve fiber of a cockroach, which has a conduction speed of ten meters per second.* That's twenty-eight million times slower than electricity running through a 12-gauge wire, but it's plenty fast enough to evade your attempts to stomp on it.

There is no axon in *P. fusi,* but there *is* an excitable membrane, and I wanted to study the linkage between its electrical excitation and the flash it produced. But in order to do that, I first needed to figure out how to piss on some fence posts.

THE CASE LAB was a large one. It occupied about one-quarter of the first floor of the multistory Bio II building, which sits on a bluff overlooking the ocean. The main lab had a warren of small rooms branching off it, most of which contained different setups for recording the electrical activity of either nerve bundles or individual neurons. These rooms were chock-full of fantastic toys: oscilloscopes, amplifiers, high-voltage power supplies, photomultiplier tubes . . . It was tech heaven. There was just one catch. As the newest member of the lab, I was at the bottom of the pecking order and, as such, eligible for only the dregs in terms of equipment and space. My fellow lab mates (all twelve of them) were super friendly and helpful, except when it came to what constituted the primary currency of the lab: the gear, which was conspicuously branded with their initials, along with warnings like TOUCH THIS AND DIE.

During his career, Jim Case produced an impressively long string of graduate students who have gone on to highly productive science careers. He attributed his success in this regard to what he called his policy of benign neglect. His students were left to their own devices to a much greater degree than occurs in most labs. We were thrown

---

*That's 22.4 miles per hour, which doesn't warrant a speeding ticket, even in Boston, which recently instituted a 25-mile-per-hour default speed limit to try to escape its title as "the most exciting city in America in which to drive."

into the deep end of the pool, where we'd either sink or swim. After showing me around on my first day in the lab, he rather offhandedly suggested that I could get started by sharing a rig with Linda, a graduate student who was working on the electrophysiology of bioluminescence in fireflies.

Her rig had all the gear I needed for the kinds of recordings I wanted to make, but sharing a rig is like sharing a car with someone who adjusts the seat and mirrors differently than you do. Linda was understandably none too pleased with the arrangement, but she managed to be gracious, so long as I could make myself invisible by working entirely around her schedule and returning all settings on the amplifiers and oscilloscope to her preferred positions, the micromanipulators to her preferred locations, the chair to her preferred height, and the microscope to her preferred focus.

Scheduling was the biggest challenge, because *P. fusi* wasn't bioluminescent anytime I wanted to work on it. It produced flashes only at night. Fortunately, I didn't have to become nocturnal, because it was possible to maintain the cultures in incubators on a reverse light/dark cycle so that the cells were fooled into being luminescent during the day, when I much preferred to work on them. However, if they were exposed to too much light during their dark phase, their bioluminescence would shut down, so I was forced to do all my microscopic manipulations under red light. This reduced their tendency to shut off their luminescence but made it very difficult for me to see what I was doing.

Also, I was finding it difficult to measure consistent flashes. They seemed highly variable. The first time I ran up against this problem, I was doing a series of measurements to test membrane excitability. I was called out of the room briefly, and when I came back, the first stimulus I applied resulted in a flash that was literally off the charts. The flash was so bright that it saturated the amplifier. You expect some recovery from fatigue, but this wasn't just a whole lot brighter; it seemed to be a whole lot faster as well, with both a more rapid onset and a shorter duration—*P. fusi* was jamming more light into a shorter period. It was so unexpected that I initially thought there might be something wrong with my light measurement system, but when I observed a cell through the microscope there was no question: That first flash was a whopper.

I began to wonder what kinds of flashes a cell would produce if it hadn't been stimulated previously within a given night phase. I didn't realize it at the time, but this was going to become a theme throughout my career in bioluminescence—how to observe the light-emitting abilities of organisms so that the act of observing didn't influence the outcome.

To run proper tests, I needed a dark room, someplace where nobody else was going to walk in and flick on the lights. Luckily, there was some lab reorganization going on that freed up a small room right off the main lab. All it needed was some way to enter and leave without letting light in. Over a weekend when the lab was largely unoccupied because of an out-of-town science meeting, David and I went to work. We constructed a light-tight antechamber just inside the door with a solid-wood frame and heavy-duty black plastic sheeting for the walls. Thanks to David's construction and design skills, it was an architectural marvel that included a sliding door on rollers that I could open and close after shutting the main door, preventing light leakage.

To provide additional insurance against the entry of unauthorized photons, I hung a flap of black plastic over the slider as a light baffle and cut it off at exactly my height. A lot of good-natured ribbing went on in the Case Lab and, at five foot two, I had come in for my share of short-people jokes, so I hung a sign on the door that read, ALL YE WHO MUST STOOP TO ENTER ARE TOO DAMN TALL. I was a little unsure how Case would respond to my grab for territory, so I was much relieved when I came in Monday morning to find that he had got there ahead of me and written a big A+ on my sign.

When Virginia Woolf extolled the benefits of having a room of one's own, she was focused on the need for carving out space and time for writing. I now had space and time for doing science on my own terms. The sense of freedom it afforded was intoxicating. I could sit in the dark with my dinoflagellates and observe their light-making one on one. It was thrilling the first time I managed to record a sequence of flashes from a cell that I knew had had no previous stimuli.

This was something that no one had ever seen before, and it was wonderfully weird. In response to a series of stimuli, the first flash was more than ten times brighter and fifteen times faster than the subsequent ones. To understand its underlying cause, I needed to be able to visualize the inner workings of the cell. Although I could see a great deal just looking through a microscope, much of what was happening with the flash was so fast, it required slow-motion play-back for careful analysis. But that would call for a high-resolution image intensifier, which was, unfortunately, exorbitantly expensive.

The funding for my research came from a modest grant that Case had from the Office of Naval Research. In those days, ONR funded basic science projects related to areas of Navy concern. Their interest in bioluminescence was strategic. The last German submarine sunk in World War I was detected non-acoustically, when it was clearly outlined by the luminescence it stimulated. In the cat-and-mouse games of anti-submarine warfare, bioluminescence had significance for both the hunter and the hunted.

ONR had been funding Case's research on fireflies in the hopes of gaining a better understanding of bioluminescence, and my research on dinoflagellates greatly heightened their interest. Case undoubtedly had exactly this response in mind when he suggested the project. The upshot was that in the next grant cycle, when he reported on my findings and requested funding to purchase a bunch of new equipment, including a high-end image intensifier, it was granted.

Now I had my dark room and all the toys I could possibly want, including a way to see what was going on inside *P. fusi*. What I found in there was like a microscopic curiosity cabinet, but rather than being dark, dusty, and filled with dead things, it was illuminated with twinkling starlight that revealed the inner workings of a cell that seemed to be constantly rearranging the furniture.

I could see that the difference between the first flash and subsequent flashes was a consequence of the summed activity of microflashes from the scintillons. The first flash was so bright and fast because the scintillons were activated in synchrony, while in subsequent flashes they were asynchronous, appearing like a scintillating star field where the microflashes increased in number to a peak and then decreased in number more slowly, sometimes with individual scintillons emitting light more than once during the course of the flash.

I was discovering more about the electrical activity that triggered the flash. Unlike most animal neurons, where the action potential is initiated by sodium ions rushing through membrane channels, in *P. fusi* it was hydrogen ions. To trigger the action potential, all I needed to do was bump the cell in a way that distorted the membrane. And I found that the action potential that triggered the light output was present in both day-phase and night-phase cells, even though no flash was produced in day phase.

The shutting down of bioluminescence was accompanied by a total rearrangement of the cell's interior. During the day, the scintillons left the periphery of the cell and migrated in to cluster around the nucleus, near the center of the cell, while the opaque chloroplasts—the organelles where photosynthesis occurs—migrated outward, spreading across the surface of the cell like a light-gathering solar array, maximizing light capture. By contrast, at night the opposite arrangement occurred, with the chloroplasts clumping around the nucleus, while the scintillons spread across the cell's surface to optimize light emission. At sunrise and sunset, the region around the nucleus was like rush hour at a futuristic spaceport, with organelles shuttling back and forth in what amounted to highly organized chaos.

These cells don't have brains, much less eyes, nor are they able to swim. They float where the water takes them. What possible reason could there be for a single cell to require such complex control mechanisms and peculiar flash patterns?

At one time, the purpose of light emission in dinoflagellates was so unfathomable that it was simply deemed to be a by-product of some other cellular function. But clearly a daily rhythm that assures that bioluminescence occurs only at night, when it will be visible, suggests otherwise.

The possibility that the bioluminescence serves a defensive function became apparent in 1972, with an elegant experiment that took advantage of the circadian rhythm in dinoflagellate luminescence to demonstrate that insect-like creatures called copepods, which are common dinoflagellate predators, grazed less on bioluminescent night-phase cells, as compared with non-bioluminescent day-phase cells. Why should bioluminescence prevent copepods from feeding?

Most copepods find their prey by setting up a feeding current, which they create by beating paintbrush-like appendages, drawing phytoplankton toward their mouths. This sounds indiscriminate, like water disappearing down a drain, but it's not. Copepods can actually manipulate their feeding currents and sort the particle stream into rejected and accepted food.

So one hypothesis is that the flash serves as a warning of toxicity, causing the copepod to reject cells that flash. Many bioluminescent dinoflagellates are toxic, and being toxic is a good way to deter predators, but only if predators recognize the prey that they need to avoid. If they have to nibble on the prey every time to determine its toxicity, then nobody wins—the prey is damaged or dies and the predator gets sick or dies. Far better for both players if the predator can learn to recognize toxic prey from a distance, through the prey's advertising its toxicity. The vibrant orange-and-black wings of monarch butterflies, which are toxic to birds, are one such advertisement. It's a very clear visual signal, effectively saying, *Don't eat me or you'll be sorry!* Perhaps the dinoflagellate flash conveys the same message.

An alternative hypothesis is that the flashing of bioluminescent dinoflagellates functions as a burglar alarm. Just the way the beeping horn and strobing lights on a car's alarm system serve to expose a burglar, forcing him to flee or risk capture, the dinoflagellate flashing might expose a copepod—which would otherwise be swathed in a protective cloak of darkness—to detection and consumption by its visual predators such as fish.

The fear screams of prey, heard in some frogs, birds, and monkeys, are examples of these putative burglar alarms. A frog being eaten by a water shrew emits screams loud enough to attract the attention of a hawk that might attack the shrew and cause it to drop the frog. The point of a burglar alarm is for prey, in the face of imminent death, to use whatever attention-grabbing means are available, such as sound, light, or odor, to attract the attention of another predator that may attack their attacker. This not only offers the prey an opportunity for escape but has the added benefit of possibly permanently removing the predator from the scene.

I have spent most of my career observing bioluminescent flashes and trying to understand what information they convey. My early experience with the strange flash patterns of *P. fusi* taught me to recognize that not all flashes are alike. There are big differences, and those differences have meaning. Although many discussions of bioluminescence in dinoflagellates assume that flashes serve the same function for all dinoflagellates, I think not. Unlike *P. fusi*, most dinoflagellates that can emit light do so in the form of one or two very dim, short flashes. Many of the dinoflagellates that emit such flashes are toxic, which suggests that, for them, the flash is a warning of toxicity. In such a case, a tiny pinprick of light in the darkness is all that is needed, because it's a private conversation between the prey and its predator. The dimness of the flash benefits both, allowing the prey to conserve energy and the predator to avoid detection by its visual predators, which means it doesn't need to stop feeding and swim away every time it sees a flash.

By contrast, the first flash of P. fusi is more than one hundred times brighter and, as such, is in no way a private conversation. It is a scream for help that will readily draw the attention of visual predators and continue to do so with its subsequent numerous and prolonged flashes, illuminating the offending attacker, making it easy prey for its predators. In the face of such exposure, it behooves the predator to stop feeding and swim away.

Nature is frequently far more complicated than we appreciate, and that complexity can be a source of confusion when we make simplifying assumptions—like the supposition that one bioluminescent dinoflagellate is like any other. There have been a number of experiments carried out over the years to test the burglar alarm hypothesis versus the toxic warning one, with sometimes conflicting results. I believe that these can best be explained by the different species and different concentrations of dinoflagellates used.

One of the dinoflagellates used in these experiments is one of those with a dim flash, *Lingulodinium polyedra*.* Oddly, while copepods refuse to eat *L. poly* in its bioluminescent night phase, they gobble it up like candy during its day phase. In other words, it's not toxic, which seems to disprove the toxic-warning hypothesis. However, it's important to remember that nature is full of cheats—organisms that send false signals. There are many examples of animals that mimic the appearance of toxic or otherwise dangerous animals in order to take advantage of predators' learned avoidance

---

* Formerly known as *Gonyaulax polyedra*. As I mentioned earlier, scientists are very picky about the naming of things. The proper assigning of names is actually the foundation of biology, but that doesn't mean that the taxonomists who come up with those names are well loved. In fact, the renaming of this particular organism raised some considerable ire in the bioluminescence community because so much research had been published on it under its old name.

behavior. The brilliant red, black, and yellow banding pattern of coral snakes advertises their toxicity to predators like hawks and coyotes. King snakes, though not toxic, masquerade as coral snakes with a similar banding pattern,* thereby acquiring protection from predators without expending the energy needed to synthesize the toxin.

Also, in some experiments with toxic dim-flashing dinoflagellates, copepods behave as they would with a burglar alarm: They stop feeding and swim away. However, in those experiments the concentration of dinoflagellates used was very high. My suspicion is that when the dinoflagellates become too abundant, even though the flashes stimulated by the copepod feeding are dim, there are so many of them that the copepod risks visual exposure and so executes an escape.

Long after my graduate student days, thanks to the experimental studies of a couple of excellent grad students in my own lab, Kathleen Cusick and Karen Hanley, working with dim emitters like *L. poly*, we were able to demonstrate that in a mixture of bioluminescent and non-bioluminescent plankton, copepods feed selectively on the non-luminescent food at low cell densities. However, above a certain threshold number of bioluminescent dinoflagellates, they stop feeding on the non-luminescent prey and swim away. On the other hand, we found that with bright emitters like *P. fusi*, a single flash is enough to attract the attention of fish predators, so its bioluminescence functions as a burglar alarm at any cell density.

Further adding to the complexity of these interactions are recent findings demonstrating that copepods release chemical cues that cause some bioluminescent dinoflagellates to increase their light output. These chemicals can also lead certain toxic dinoflagellates to

---

* It's not a perfect disguise, which is why learning this poem can save your life: "Red touch yellow—kills a fellow. Red touch black—venom lack." On the other hand, if you're not into poetry, just try to remember that the one with the black nose is bad news.

increase toxin production. In other words, dinoflagellates can sense the presence of predators and adjust their defenses accordingly. With every new discovery, the world of sea sparkle becomes "curiouser and curiouser!" *

Relationships between living creatures are extraordinarily complex. Trying to sort out these associations to determine what part different bioluminescent signals play in a laboratory environment is fraught with pitfalls. It's hard enough to keep one critter happy and healthy in a laboratory setting; when you're dealing with three or more, in order to sort out multitrophic connections, the challenges quickly multiply.

To comprehend the meaning of any animal's visual signals, it's essential to put yourself in their shoes—or chelae or flippers. In other words, you need to be able to imagine what their world looks like. That's a grand challenge, because even though it's part of our world, it's a place very few people have seen in its natural state.

* Almost as trippy as *Alice in Wonderland*!

Chapter 4

THE STARS BELOW

The knot in my stomach tightened as I walked out onto the fantail to look yet again at the steel cable disappearing into the depths behind the ship. Dangling from the end of that cable, some eight hundred feet below, was friend and fellow scientist José Torres, taking his turn in a deep-sea diving suit called Wasp. The sun was getting low in the sky and I was starting to worry that I might have to wait until the next day for my first deep dive. The Wasp crew said they were saving the best for last, but I wasn't fooled: I knew they were saving the shortest for last.

Developed by the offshore oil industry for the purpose of diving on oil rigs to depths of two thousand feet, the suit is called Wasp because it looks like a big yellow insect with a transparent head, a yellow tube body, and metal Michelin Man arms with pincers instead of hands. There are no legs for walking on the bottom. Instead, the suit has thrusters on the outside, controlled by foot switches attached on the inside to a metal plate that can be adjusted to accommodate the stature of the occupant. It took time to pump up the floor; consequently, going in order of height and saving the shortest for last made sense, but for me it felt like that interminable stretch of time leading up to Christmas, when, as a kid, the more desperately I wanted it to come, the more drawn out its advent became.

It had all started innocently enough, three years earlier, in 1981, when Dr. Case scored a new toy for the lab. The Optical Multichannel Analyzer (OMA) was the most sensitive spectrometer ever developed for measuring the color of dim, transient light sources. From the moment it came into the lab, I couldn't keep my hands off it.

Traditionally, color is measured by dispersing light with a prism or a diffraction grating, either of which will transform white light into a rainbow. The prism or grating is rotated so the rainbow is scanned across a sensitive light detector that measures the spectrum, one narrow bit of the rainbow after the next. This works well for measuring spectra of relatively stable light sources, but not for measuring brief flashes of bioluminescence. The OMA eliminated the need for scanning by replacing the single photomultiplier tube with a row of seven hundred solid-state detectors that could measure the contributions of different parts of the rainbow all at once instead of consecutively. It was an awesome technological breakthrough that opened up a whole new way of looking at bioluminescence.

Color has all kinds of interesting stories to tell when viewed with spectrometers, which are not subject to the kinds of deceptions to which our eyes may succumb. As I had experienced with those yellow roses in the hospital, the brain cannot be trusted to report colors accurately. When the light is too dim for the red, green, and blue cones that are the basis for our color vision, then our more sensitive rods take over, reporting light but not discriminating color. Nonetheless, the brain may supply the sensation of a particular color, informed by expectations rather than actual data.*

Even when the color is bright enough for the cones to detect, they can be fooled, since pure green light can look the same to our eyes as a mixture of blue and yellow light; and what we call white light can be any of a wide variety of color combinations that stimulate our three different cone types equally. Color in the ocean carries important in-

---

*Much like modern politics.

formation about how animals are seen (or not seen) and what their light-emitting abilities may have evolved to communicate. Details related to the relative contributions of specific colors, plotted as spectra, can also provide hints about the chemicals responsible for producing the light. The OMA promised exciting new discoveries, if only there were some way to gain access to living deep-sea light emitters so we could measure their bioluminescent emission spectra.

As chance would have it, there was, because just down the hall from Jim Case's lab was Jim Childress's. Childress, a pioneer in the study of metabolism in deep-sea animals, had perfected methods of bringing the animals up alive. Despite a common misperception that deep-sea dwellers die from extreme changes in pressure when brought to the surface, changes in temperature are far more damaging. Deep ocean water, which constitutes about 90 percent of the volume of the ocean, is very cold, averaging between thirty-two and thirty-seven degrees Fahrenheit. If you capture deep-sea animals in a net and drag them up through warm surface waters, they are basically cooked and consequently either dead or dying by the time they reach the surface. Pressure changes are less damaging for many of these animals because they don't have air-filled spaces, like air bladders, and thus don't experience explosive changes in volume. Childress was able to keep animals alive by capturing them with nets that directed the catch into thermally insulated capture devices of his own design.

Usually, at the end of the net there is just a mesh bag, called the cod end,* that holds the catch. Childress replaced that bag with a large-diameter PVC tube that contained a mesh bag inside to retain

---

*Although an argument can be made that a cod end and a codpiece share a similar shape and therefore a possible common word origin, I can find no connection. *Cod end* refers to the end of the net where the codfish accumulated, while *codpiece* derives from the Middle English *cod,* meaning "scrotum," and refers to a male fashion accessory that has largely fallen out of favor except with certain comic-book superheroes.

the catch, and ball valves at either end of the tube that could be closed at depth, sealing the catch in the ambient cold water. In this way, animals could be brought up alive, and if they were kept cold aboard the ship, they could usually be maintained long enough for Childress and his team of grad students to make critical measurements related to their metabolism and, Case reasoned, long enough for us to study their bioluminescence.

Thus it came to pass that I went on my first ocean expedition in 1982, aboard the 110-foot research vessel *Velero IV*. One hundred and ten feet is one and three-quarters bowling alleys long, which is not a lot. One bowling alley length housed the bridge, the mess, the lab, and the sleeping quarters for eleven scientists, seven crew members, and one ship's cat, Buffy, while the remaining three-quarters was the fantail or back deck. The dry lab* was in the bowels of the ship and required maneuvering all the scientific gear down a steep ladder and then strapping absolutely everything down so it wouldn't go flying the first time we hit rough seas. The mess was tiny, the food terrible. The four-person cabin I was in was cramped and musty. Opportunities for sleep were rare and usually uncomfortable. I had a top bunk that was crosswise to the ship's main axis, so when the ship rolled side-on to the waves, I slid up and down the length of the bed. In rough seas, the porthole at the foot of the bunk leaked and the mattress grew soggy. Nine scientists shared one head with sketchy plumbing. Those in the know maneuvered to be between third and fifth in line for the shower—the sweet spot between when the hot water first surfaced and when there was no more. For anything other than liquid, the toilet required four or more flushes—a common source of aggravation. I once saw the head door slammed open in a fit of pique as one of my fellow grad students emerged,

---

*Most shipboard research facilities have a wet lab and a dry lab. The wet lab is kind of like a mudroom in a house, where all the messy, wet stuff is dealt with, and the dry lab is where the more sensitive electronic gear is set up.

holding a turd in his bare hand. He flung it over the side of the ship and then, seeing me, muttered, "Fucking floaters!"

Going to sea on research expeditions is not for everyone, but I loved it. This was the kind of swashbuckling adventure I had dreamed about as a kid. A two A.M. shout into the cabin of "Net's up!" would send us all scrambling out of our bunks and into our wet-weather gear. Out on deck, we each took up our assigned position or task, working the hydraulic A-frame, crewing the capstans used to haul in the tag lines, schlepping five-gallon plastic carboys of seawater from the chiller to set at the ready next to the trawl bucket. We each had a vital role to play, and we depended on and looked out for one another. If you screwed up and weren't paying attention as the heavily weighted net appeared over the transom and went sliding across the deck, somebody could get seriously injured. It was a brief period of high stress, rewarded by the extraordinary payout of whatever came whooshing out of the cod end as it was held over a large metal washtub and the ball valves were cracked open.

During night recoveries, the contents of the trawl bucket glowed with streamers of liquid blue light: scintillating plankton, glowing krill, and mangled but occasionally still-pulsating jellyfish. When the trawl bucket was carried into the wet lab, everyone congregated around it, plunging bare hands into the bone-chilling water to haul out one animal after another. I joined in, pulling out a bright red shrimp* the size of a hamster. It had long red antennae, a beautifully sculpted carapace with elegant curved spines, and a multitude of feathery legs, and when I lifted it out of the water it spewed brilliant streams of sapphire-blue light from nozzles on either side of its mouth. The light pooled in my palm and spilled between my fingers, dripping back into the bucket, where it continued to glow. Also in the bucket were lanternfish, each sporting light organs called photophores that looked like jeweled studs adorning their sides. There

---

* Gnathophausia ingens.

were hatchetfish, so named because their body shape is like a hatchet; its tail forms the handle and its silver-sided body the blade, and along the bottom edge of that blade are two rows of light organs that look like two-toned fingernails painted mostly silver with magenta lunules.

Besides these more common species, it seemed like every haul held at least one special surprise. There was a velvet-black dragonfish, slender and long, like an eel with a whiplike bioluminescent fishing lure protruding from its chin. There was a vampire squid, an inky black, gothlike creature with eight arms connected to one another by a web of skin and each lined with fleshy spikes. It had two enormous eyes at the base of the arms and two large, lidded light organs that looked like a second set of eyes at the base of two big flapping fins. There was even a stoplight fish, a midnight-black beauty with a big red light organ under each eye and a smaller blue one right behind it.

As it turns out, bioluminescence comes in all colors: red, orange, yellow, green, blue, and violet. In the open ocean, blue dominates. This makes sense if you think in terms of efficient visual communication. The reason everything looks blue underwater is that blue is the color that travels farthest through water. The other colors are scattered and absorbed to varying degrees and gradually disappear. You may notice the very weak penetration of red light if your scuba buddy is wearing something red. Above water, a red Speedo appears red because it absorbs all colors except red light, which reflects off the bathing suit and back to your eyes. Deep enough below the surface, however, where there are no more red photons, the suit absorbs all available light and appears black.*

It seems wrong that the color of a thing is defined by a negative—in other words, what it *doesn't* absorb. Chlorophyll appears green because it absorbs red and blue, using the energy from these colors to

---

*And hopefully more slimming.

make photosynthesis possible. The green that reflects back to our eyes is the unuseful stuff—basically discarded photons. Most of the visual information we take in is in the form of rejected photons—that is, reflected light. Bioluminescence is an exception to this general rule because it is emitted photons. That so much of this bioluminescence is blue helps explain why so many deep-sea animals are red: If the only light to see with is blue, being red is akin to being black. Red pigments absorb blue photons, reflecting nothing back to the eyes of predators.

Since downwelling sunlight filtered through seawater is blue and the majority of bioluminescence is blue, most deep-sea animals have evolved eyes that see only blue light. The stoplight fish is different from most of its kind. It sees blue light, but it can also see red light, which means it's got sniper-scope vision! To be able to see and sneak up on prey undetected is a superpower with the added benefit that red light helps break the camouflage of one of the stoplight fish's common prey items, red shrimp. While a red shrimp in blue light appears black and well camouflaged, under red light it will stand out like a beacon in the darkness. And there is potentially still one more benefit: This remarkable fish can use its red light at close range to communicate with a prospective mate over a private wave band without fear of attracting the attention of visual predators.

I had read about many of these animals. I had seen pictures of specimens preserved in formalin and pencil drawings of what they were supposed to look like. I had read statistics on how the vast majority of animals brought up in trawl nets are bioluminescent. But I was still gobsmacked by the reality of so many fantastically strange creatures with multiplicities of light-producing means and methods.

In every trawl, there were examples of the different proposed functions of bioluminescence that I had been reading about. There were lights for finding food, either in the form of bioluminescent fishing lures used to attract prey or built-in flashlights for locating prey in the dark. There were lights for attracting mates with different-shaped light organs or different flash patterns, used as species and sex identifiers. And there were bioluminescent defense strategies, like spewing luminescence into the water to distract a predator, or very bright light organs on the tails of some lanternfish that could be used to temporarily blind a pursuer. And virtually all the fish, shrimp, and squid sported belly lights, used to eliminate the silhouette that is the search image of so many open-ocean predators. This form of camouflage is so commonplace in the ocean that it's the equivalent of color-matching camouflage on land.

One of the fish I pulled out of the trawl bucket was a saber-toothed viperfish, its awesome name a consequence of the fearsome curved fangs that protrude from its lower jaw. These teeth are so long and so sharp that if they closed inside the mouth, they would impale the fish's own brain. Instead, they slide into grooves in the upper lip, and when the mouth is closed they extend to a point just above the eye. That should be more than enough badassery for any one creature, but this fish piles on with a panoply of light organs that stretch the imagination in terms of possible functions. An elegant long fin ray grows out of its back and arches forward, dangling a luminescent lure in front of its fearsome maw. Clearly, this must be used for fishing. Two rows of prominent photophores along the belly certainly serve to camouflage the fish's silhouette, hiding it from the eyes of upward-looking predators. A flashlight under each eye might aid in finding food or attracting a mate. But what of the photophores inside the mouth? Are they another means of attracting prey, or do they illuminate the long, lucent fangs, perhaps as a threat? Even more bizarre are the tiny, inconspicuous light organs embedded in a mucous layer that covers the back, belly, and fins. When these organs light up, the fish strobes an outline of its body—for what? Defense? Sex? Disco dancing?

Another unexpected prize from the trawl was an exceptionally large anglerfish. Most deep-sea fish are small—an adaptation to living in a food-poor environment. Hatchetfish are the size of a silver dollar. Lanternfish are no bigger than a pocketknife. Even the fearsome viperfish is generally less than a foot long. The apparent ferocity of anglerfish is often much diminished in viewers' eyes when they learn that gruesome countenance is associated with a fish the size of a plum, or in some cases a plum pit. But this anglerfish was the size of an eggplant—a big one. Like most anglers, she had an enormous mouth filled with needle-sharp teeth and a bioluminescent lure called an esca. But this lure looked like it was designed by Dr. Seuss. It consisted of a short, stout rod protruding from her upper lip and crowned by a tulip-shaped light organ, festooned with two bundles of long, delicate translucent threads. Was this elaborate structure an adaptation for attracting prey or mates? Both are possible; some lures seem to mimic small prey, while the ornateness of others is believed to aid a male in identifying a female of his own species.

The male anglerfish is much smaller than his female counterpart. He lacks a lure and has no teeth for consuming prey. For many anglerfish species, the male's only hope for continued existence is as a gigolo. In the unimaginably immense black void of the deep sea, he must somehow locate a potential mate, either visually or by smell, and, upon finding her, seal the relationship with an eternal kiss by latching on to her flank, where his flesh fuses with hers. Her bloodstream then grows into his body, providing him with sustenance, in return for which he provides sperm upon demand. This lifetime commitment may sound romantic, but it's not all hearts, flowers, and pillow talk. He's a bloodsucker and a sperm bag, and she's ugly and weighs half a million times more than he does.*

And she has a nasty temper to boot. I witnessed her vicious streak when I had her in an aquarium and tried to photograph her head-on to capture the full measure of her stunning unattractiveness. I was

---

*This describes the most extreme example of sexual dimorphism in anglerfish, found in the northern giant seadevil, one of the largest species, with the female achieving gargantuan proportions of up to 3.9 feet long and further adding to her allure by being covered in warts.

using a long-handled paintbrush to occasionally nudge her back end around so she would face the front of the aquarium. Every time I touched her side, no matter how gently, she would twist around and snap at the point of contact. I presume a male might expect the same reception, which suggests he may need to execute great caution in selecting his method of approach and point of attachment.*

When the expedition was over, I had a hard time readjusting to life on land. It wasn't obvious why. I was pretty sure I wasn't missing the sleep deprivation, bad food, or lousy plumbing. Eventually, though, I realized that what I did miss was the excitement and camaraderie of being at sea. Still jacked up on adrenaline, I felt the world was off kilter. All the students roaming around campus seemed alien and clueless, nothing like the members of the tight little team I had been part of at sea, and I was dismayed by their obliviousness to the secret world revealed by our net hauls. How was it possible that they didn't know that just a hop, skip, and deep dive offshore there are outlandish life-forms festooned with headlights, taillights, belly lights, mouth lights, fishing lures, and light-spewing nozzles? That should be front-page news—right? That nasty-tempered angler we brought up turned out to be a species never seen before! Why was that not a banner headline? It was incomprehensible to me that the world held such wonders, about which most people knew almost nothing.

WHILE MANY MAY claim that getting a Ph.D. is as easy as riding a bike . . . through a desert, with no sleep, while people in black robes try to distract you by setting your hair on fire . . . that was not my experience. I loved the five years I spent completing my degree. It was without question the best academic experience of my life. Which is why, when I graduated in 1982 and it came time to move on, I was finding it difficult to get excited about what should have been a great opportunity.

---

* Good advice for anyone when it comes to mating.

A couple of months after passing my orals, I flew from sunny Santa Barbara to cold and sunless Madison, Wisconsin,* to interview for a postdoc position in the laboratory of a scientist on the leading edge of research in excitable membranes. The interview went well, and I was offered the position. I tried to feel enthusiastic, but the truth was, I was disheartened by the idea of moving so far from the ocean. I believed I *belonged* on a ship, and leaving just felt wrong. Jacques Cousteau claimed, "The sea, once it casts its spell, holds one in its net of wonder forever." I was in the sea's thrall. At least I had one more research expedition to look forward to before my departure, and it promised to be especially exciting, because it involved a new approach to deep-sea exploration.

Bruce Robison, at that time an associate research scientist at UCSB, had spent years going to sea to study midwater fish on the kinds of expeditions I had just experienced. Then, one fateful day, he and Jim Childress were walking across campus and spotted a sign for FREE DONUTS AND COFFEE attached to an announcement for an ocean engineering seminar. They figured they'd check it out. In addition to the caffeine-and-sugar rush, they were treated to a movie about the Wasp deep-sea diving suit and a post-film discussion of its engineering functions. Robison, known as Robi to his friends, wondered about possible science applications. He was frustrated by the limitations imposed by net sampling and wanted more direct access to what constituted the most unexplored frontier on our planet, the midwater.

THE FIRST SUBMERSIBLE to carry humans into the deep sea did so in the early 1930s, when William Beebe, a scientist with the New York Zoological Society, and Otis Barton, an engineer, made a series of thirty-five deep dives (maximum depth, 3,028 feet) off Bermuda

---

* I'm aware that Madison has sunshine. It just didn't appear at any time while I was there.

in a steel sphere of Barton's design. Dangling from a steel cable, the 5,400-pound sphere and its two occupants were hoisted up and down through the water by a steam-powered winch while Beebe, peering out through a six-inch porthole, made observations that he later included in a series of articles for *National Geographic* and in a book entitled *Half Mile Down.*

Beebe was a gifted raconteur, and his words opened a portal to a heretofore alien world. Besides inspiring future explorers and environmentalists, including E. O. Wilson, Rachel Carson, Jane Goodall, and Sylvia Earle, he is also credited with helping to pioneer the field of ecology, championing the need to study animals in their native habitats.

After those record-breaking dives, there were numerous advances in submersible technology, but the emphasis was on vessels designed for exploring the seafloor, where geologists could collect rocks and biologists could find animals, like deep-sea corals, that didn't run away. Most ocean explorers viewed the midwater as a wasteland that had to be traversed to reach the interesting stuff on the bottom. The fact that net sampling and Beebe's observations suggested otherwise was largely unappreciated, which is why Robi's idea of adopting Wasp as a tool for exploring the midwater was not readily embraced. He had a tough time getting the project funded, but he persisted, and this expedition, which took place in the fall of 1982, was the payoff.

This was real pioneering stuff and I was going to be part of it, albeit in an ancillary role. I was invited along to measure the bioluminescent emission spectra of the midwater animals, especially any fragile deepwater jellies that they managed to capture with the Wasp.

It was thrilling to be included, but since I wasn't trained to pilot the suit, I had to satisfy my curiosity by listening to verbal descriptions provided by those who were. Their work was focused primarily on observing and capturing animals with the Wasp's lights on, but whenever I could, I would get on the headset and ask whoever was in the suit to turn out the lights and tell me exactly what they saw. Their scientific reports amounted mostly to exclamations of "Oh, wow! That is so cool!" I would beg them to be more specific but found their attempts less than satisfying.

Mostly, I was left with the knowledge that there was lots of light, but it was not bright enough to film with the underwater camera mounted on the Wasp. The only way to observe it was to go see for myself, which I couldn't yet do. It was torture. Robi took pity on my obvious frustration and told me that in two years (1984) they were planning a second Wasp expedition, and if I was still around, maybe I could get trained as a pilot and go see for myself.

It was a phenomenal offer, but to turn down a plum postdoc in favor of the road less traveled—actually, the road *not yet* traveled— seemed insane. If I did this, I'd have no plan B. But, for the first time since getting out of the hospital, I didn't care. The only thing that mattered to me was seeing what was down there, risks and consequences be damned!

AND SO HERE it was two years later. In the interim, Jim Case had hired me as a postdoc so I could finish up the spectrometer work with the OMA. I had taken it on four major trawling expeditions— two off California, one off Hawaii, and one off the northwest coast of Africa. I so loved what I was doing, seeing and learning about the bioluminescence capabilities of all these amazing animals, that the two years had flown by. But now time seemed to have slowed to a stop, and I was on tenterhooks waiting for my first deep dive.

The goal for this initial day at sea was to accomplish the final stage of our instruction by giving each of the five of us who had been trained as pilots the experience of being lowered into the depths. This wasn't so much a tech check as a psych check. Scooting around the fifteen-foot-deep sunlit tank we had trained in at Port Hueneme was a very different experience than being dropped through hundreds of feet of water into the dark, cold depths of the vast ocean, encased in an ill-fitting metal sheath. If anyone was going to have a claustrophobic meltdown, the Oceaneering team in charge of the Wasp wanted to know at the outset.

Oceaneering was the company that owned and operated the Wasp. Charlie Sandstrom, a weathered sea dog with many years of experience working around ships and oil fields, was their man in charge of this odd little endeavor. The other members of his team were young guys with backgrounds in diving, underwater salvage, and oil-rig work. Their mission was to keep the Wasp in working order and to try not to get any of the scientists killed. The scientists' mission was to test the value of the suit as a tool for ocean exploration.

Most of us live our lives with our feet planted firmly on the ground and are out of touch with the true nature of our watery world. Only 29 percent of the surface area of Earth is land; the rest is water. Oceanographers often try to alert their fellow Earthlings to the significance of their research by using such numbers. Or they describe how we have better maps of the back side of the moon than we do of the bottom of the ocean. But even that disparity misses the point, because it's thinking in just two dimensions. The three-dimensional reality is that, while living space on land extends into the tallest trees and many feet beneath the surface, it is still an absurdly thin layer and represents a tiny volume compared with the staggering enormity of the ocean, which encompasses more than 99.5 percent of the living space on the planet. And this is no empty void. These waters teem with life, but our experience of this largest of Earth's ecosystems is scant and biased by the limitations of our tools for exploration.

It's remarkable that the primary way we know about life in the midwater is by dragging nets behind ships. How many other branches of science can you name that still depend on technology thousands of years old to gather data? It's an axiom among marine biologists that nets capture only the slow, the stupid, and the greedy:

Some animals swim too slowly to get out of the way. Some don't recognize the approaching net as a threat, but then, why should they? And still others are predators that dart into the net to grab what looks like an easy meal but don't make it out and are therefore victims of their own gluttony. But how many are there that we don't know about because they are simply too fast or too wily to be captured by such primitive means? And, aside from all the animals our nets miss, there are also the fragile, gelatinous forms that they shred. Until we were able to enter their world and observe them directly, scientists had no clue that the gelatinous goo seen in our trawl buckets came from a myriad of diaphanous life-forms that had been mauled beyond recognition.

My very limited understanding of life in the midwater was based on net-captured creatures. Observing them in the trawl bucket was like sifting through the remains of some ancient civilization and trying to imagine what the daily lives of the people of that world were like. Only now I wasn't going to have to just imagine. I was about to see for myself—if José would just stop screwing around down there.

Just as I was beginning to think that Christmas would never come, I heard the whine of the hydraulics kicking on and saw the cable begin to rise. Robi came to join me on the fantail and watch the Wasp's winch reel in the two-thirds-of-an-inch-thick steel-reinforced cable that carried power and communications between the ship and the suit. Known as the "umbilical" or "tether," it attached to the top of the suit just above and behind the transparent acrylic hemisphere that was the Wasp's view port.

At the moment when the top of the Wasp came into view, some sixty feet below the surface, a crew member who had been leaning over the back of the fantail to watch for it turned toward Charlie at the winch control console and used two fingers to point to his own eyes, the sign for "in sight." The winch stopped and two scuba divers flipped backward out of the ship's tender that had been drifting off the stern. They swam down to the Wasp and attached two snubber lines that were needed to assure that the two-thousand-pound suit didn't swing on the end of its cable and smash into the back of the ship as it emerged from the water.

Once the divers were clear, we heard the winch start up again. As the suit appeared and the view port came even with the fantail, we could see José grinning back at us. Clearly, he had passed the psych test. As soon as the Wasp was lowered onto the metal frame that served as its docking station on deck, José was released from the confines of the suit, and he immediately clambered out, shouting, "That was awesome!"

As Robi and the others surrounded him to hear details of what he had seen, I began my pre-dive checks. I would have loved to hear what José had to say, but, besides being anxious to see for myself what was down there, I also bore the onus of being the only female *and* the youngest member of the team; I needed to prove I was up to the task, and I sure didn't want to piss off Charlie by keeping him waiting.

Once the external inspection was complete and the floor had been pumped up to accommodate my height, I was given the go-ahead to enter. Retracing José's exit path, I climbed up the stepladder, sat on the plywood that covered the acrylic hemisphere, swung my legs into the neck of the suit, and then twisted around to face forward, resting my hands on the metal arms where they attached just below the neck and supporting my weight with my hands and my toes, which I slid down the inside front of the suit until I made contact with the floor.

I went through the internal checks carefully but as quickly as I could, calling out gauge pressures, testing the thrusters, confirming the presence and functionality of all safety and emergency gear. We had been thoroughly drilled in emergency procedures during our training, so I was cognizant of the life-threatening implications of all this paraphernalia, but I was already fully committed to this adventure. No turning back now.

It was time to dive. I gave the thumbs-up and a crewman swung the view port up and locked it, sealing me inside. The instant the hatch closed, all the engine noise, which follows you everywhere aboard a research vessel, was muted and replaced by the soft whir of the scrubber fans. Throughout the dive, I would breathe air at atmospheric pressure, which meant that despite the immense depths to which I was descending, there would be no need for special gas mixtures, no danger of nitrogen narcosis or decompression sickness ("the bends"), and no need for decompression afterwards. An oxygen tank bled a slow stream of oxygen in while the scrubber fans pulled the air from inside the suit through an absorbent material that chemically extracted the CO_2 I was exhaling. The ability of the pressure hull to keep the internal pressure at close to one atmosphere, even as outside pressures exceeded hundreds of pounds per square inch, is what qualified Wasp as an atmospheric diving suit, or ADS.

I felt the suit rock back as the winch lifted it clear of its support stand. The A-frame hoist rotated out, carrying me over the back of the ship, and as the winch paid out, I got a close-up view of the ship's stern, with its name and port of registry—T. G. THOMPSON SEATTLE— displayed in black capital letters. The ship operated out of the University of Washington, but for this mission it would be plying the ocean off Santa Barbara, between the Channel Islands and the mainland. The islands offered some protection from bad weather, and since the channel waters were generally less than two thousand feet deep, the thinking was that if for some awful reason the suit became detached from the umbilical, it couldn't sink below its operational depth limit. I was going only to eight hundred feet for this first dive—nowhere near the bottom—but for somebody who had never been deeper than ninety feet while scuba diving, that seemed like the abyss.

As the Wasp hit the water, I looked up and saw Robi staring down at me over the back of the ship; then water closed over the dome and he disappeared from view. As I had been trained to do, I checked for any sign of leaks around the perimeter of the view port, and when I saw none I called up to Charlie on the comms link, "I have a seal." At thirty feet, Charlie braked the winch and put me through another obligatory checklist of gauge readings and a test of the through-water emergency communications system. Once that was completed, he had the snubber lines pulled free and then reengaged the winch, and I resumed my descent.

I pulled my arms out of the suit's armored sleeves and rested on my forearms, my body tilted forward against the front of the suit. The articulating arms of the Wasp are hydraulically assisted, which makes it sound like I had cyborg superpowers, but the Wasp's arms were mere decorations as far as I was concerned. I had had to lift weights for a year just to pass my pilot's test. In the tank, I managed to make them work just enough that I could do up a shackle. Beyond that, they were useless. As the weight of the water increased, the arms became stiffer and less movable. We had been told to keep flexing them if we wanted them to stay operational, but operational for what, exactly? Oceaneering divers working on oil rigs needed to be able to use wrenches and turn valves; we had no such requirements. First and foremost, we were there to observe. The various scientific instruments that we had brought with us were designed to operate electronically, controlled by switches inside the suit. I let the arms lock up.

My nervousness subsided as I began to focus on the view outside. I felt no hint of claustrophobia. Compared with the guys, I had room to spare, and besides, all my attention was directed outward. With my face positioned in the center of the transparent observation dome, I felt completely immersed in the surrounding water. I was not thinking about the oxygen or CO_2 levels, and it wasn't cold enough yet to be bothered by the temperature. I was entirely absorbed in trying to assimilate what I was seeing.

As a scuba diver, my attention was always on the underwater

landscape—rock outcroppings, seagrass meadows, kelp forests, or coral reefs and their associated flora and fauna—but now I felt my focus shifting to a very different kind of habitat: water, as far as my eyes could see in every direction. In the midwater, there are no surfaces for organisms to settle on. This is a world of gradients. Light, color, temperature, salinity, pressure, oxygen—they were all changing as I descended. According to the readout on the pressure gauge, the grip of the pressure monster was increasing by one atmosphere (14.7 pounds per square inch) for every thirty-three feet of descent, a squeeze that I was blissfully unaware of inside my protective shell. What I *was* aware of was the changes in the color and intensity of the sunlight from above.

On first passing through the boundary between air and water, there is an abrupt change in color, shifting from multicolored to predominantly blue-green. I was familiar with that shift from scuba diving, but now I was dropping at a speed that I could never have achieved with scuba, where changing pressure would have required distracting ear-clearing gambits to prevent a ruptured eardrum.* The color was shifting from aquamarine to a grayish cobalt blue, and coincident with that transition was a change in light intensity. Measured with a light meter, this change was even more dramatic than the color change; light diminishes exponentially with depth.

In the very clearest ocean water, sunlight decreases tenfold for approximately every 250 feet of descent, but these waters weren't crystal clear, and so it was diminishing even faster because of the absorption and scattering by particles and dissolved organic matter

---

* Despite the claims of English scientist J. B. S. Haldane that a perforated eardrum allowed him to "blow tobacco smoke out of the ear in question, which is a social accomplishment," it is a condition that most scuba divers prefer to avoid. All manner of contortionist maneuvers may be employed, such as the Toynbee maneuver, the Edmonds technique, the Lowry technique, and the Frenzel maneuver, which all have the goal of forcing open your Eustachian tubes to allow higher-pressure air to pass through benignly.

in the water. Oddly, though, that's not how it appeared. One of the remarkable things about eyes is that they are not linear, but rather logarithmic, sensors. In order to allow us to see over an enormous dynamic range, from brilliant sunlight at high noon to the dimmest of starlight on a forest floor, the eye's measurement scale is compressed. It is a phenomenal capability, but to accomplish it, our eyes must lie to us, reporting a tenfold decrease in light intensity as a mere halving of available light. Consequently, I wasn't conscious of how much the light had diminished until the Wasp's small floodlights, called Snooperettes, started illuminating particles in the water. The lights had been on since the surface, but the sunlight was too bright for me to notice. Everything still looked blue overhead, but directly in front of me, beyond the penetration of my Snooperettes, it was now gray grading to black below.

I was just noting these changes when the suit descended into a layer of what looked like small red crabs. Crabs? They're supposed to live on the ocean floor. I checked the depth gauge: I was at two hundred feet—nowhere near the bottom. The fact that this was more than twice as deep as I'd ever been in my life didn't even register as I tried to wrap my head around the spectacle of all those crabs. The ones near me appeared red because of the reflected light from my Snooperettes, but those farther away looked gray—well matched to the gray light field behind them. There were hundreds of them—hanging there like an invading army of drones spaced a yard or two apart, above, below, and to the sides of the Wasp. But they weren't expending energy to hover, as drones would; they were floating with claws splayed out, mostly not moving until one would tail-flip backward briefly before returning to float mode. I had never seen these guys while scuba diving or in net hauls. Was it normal for them to be here like this, and in such numbers?

I later learned that the presence of this crab swarm, which we saw routinely in our first few weeks of the expedition, was not common. *Pleuroncodes planipes,* also known as langostino or tuna crabs, are normally found in warm waters south of San Diego, but this was an El Niño year, characterized by warmer-than-normal sea surface temperatures, which had apparently extended their range to the north.

As I continued my descent, the crabs disappeared and were replaced by more amazing creatures. There was a gooseberry comb jellyfish (*Pleurobrachia bachei*) with two incredibly long tentacles, each sporting a gauzy splay of hairlike tentilla that it rapidly retracted as it swam away after being disturbed by my bulky presence. A siphonophore called *Nanomia* swam downward, pulsing its transparent corncob of swimming bells while trailing a ropy tentacle tail. The speed and agility of these fragile manifestations of organized water was stunning, and their alternative modes of locomotion bizarre. The comb jelly moved by means of eight rows of paddles—called combs—beating in traveling waves that pushed the jelly forward or backward with equal ease, while the siphonophore pumped water from closely packed swimming bells that it contracted synchronously when moving in a straight line and asynchronously when making a turn. A little deeper, I spotted a six-inch silver-sided fish hanging vertically, head up. *What is that about?* I began to wonder how much of the animal behavior I was observing was influenced by my presence.

When I reached eight hundred feet, Charlie stopped paying out the umbilical and then briefly distracted me from my observations as he had me call off gauge readings to him. The suit oscillated slowly up and down like a tea bag on a string. For this, my first deep dive, the umbilical was a source of comfort, connecting me acoustically to Charlie and physically to the mother ship, which rocked in a gentle swell at the surface. Aside from power and communication, it provided the reassuring thought that, if anything went terribly wrong, I could be quickly hauled back from the abyss. But it also restricted my freedom to explore, as I was about to experience.

As I logged off comms with Charlie, a large jellyfish pulsed into my lights. Arched white tentacles studded the rim of a glasslike cen-

tral disk. As it swam out of the light and into the enveloping darkness, I attempted to give chase, pressing forward on the foot switches
that activated the thrusters. My pursuit was abruptly terminated,
though, as the umbilical brought me up short, like a dog on a leash.
I hadn't adjusted my ballast.

I cracked open a valve, letting compressed air into the ballast
tank. At the same time, I called up to the surface to let Charlie know
what I was doing and he cautioned me to keep an eye on the umbilical. During training, he had made it abundantly clear that not
being aware of where your tether was at all times was a major no-no.
He had reinforced this admonition with a graphic description of the
umbilical looping under the suit's arm and ripping it off, with, of
course, lethal consequences for the diver inside. I was pretty sure the
story was apocryphal, but that didn't lessen its impact. At the first
sign of slack in the cable, I closed the valve and returned my attention to the view outside.

The jellyfish was long gone. In its place were large, elegant
shrimp that looked like they were on skis and gliding through a perpetual snowstorm. The shrimp were a type of deep-sea prawn called
a sergestid (*Sergestes similis*), and their "skis" were their impressively
long antennae that stuck out and down in front and then bent back
sharply, trailing to either side of their beating pleopods. The snow
was marine snow, the white flocculent material, made up of de-
composing plankton and fecal pellets (a.k.a. poop), that rains down
from surface waters. The procession of shrimp was mesmerizing—
completely unlike the damaged specimens I had seen in trawl nets.
Each had a body that looked like it was fashioned out of cut glass—
crystal clear except for the section just behind the head, which was
cherry red. I longed to keep watching, but there wasn't time—Charlie
wanted to finish up and go to dinner. If I was going to see the thing
that had brought me here in the first place, I needed to act quickly.

It takes more than twenty minutes for the human eye to dark
adapt, so I turned out the lights, prepared to wait a while before

being able to see much, but no delay was necessary. Instantly I was engulfed in what looked like a field of stars. Everywhere I looked, there were glowing motes. The density was like what you might see in a desert sky on a moonless night, but these stars weren't static; they were swirling all around me like a three-dimensional version of Van Gogh's *Starry Night*. My breath caught in my throat.

It was difficult to focus on any one star long enough to fully see it, but after a while I realized that these weren't just discrete points of light. Much of it looked like internally illuminated organic proto-plasm that, upon closer inspection, consisted of strings of two to four tiny glowing orbs held together in a chain by a gossamer sheath. The phrase "mermaids' tears" came to mind. Their light was not a steady glow or an abrupt flash, but rather a slow, deliberate illumina-tion, like turning up lights with a dimmer switch, only these lights didn't switch on synchronously, but in a propagated sequence. I tried to time them to see how long they stayed on, but they kept dis-appearing from my field of view before the light faded out.

It was hard to be sure, but it seemed as though the tears were being stimulated by the movement of the Wasp. The motion of the ship bobbing in the waves at the surface was transmitted down the tether to the suit, causing it to undulate up and down, creating an enveloping sparkly halo. It was my impression that the sheer stresses caused by that motion were activating the tears. Occasionally, one of the strings of tears would make direct contact with the observation dome and the orbs would brighten as the gossamer sheath holding them together stretched, and stretched some more, and the lights smeared and disappeared. Mixed in with all these mermaids' tears were other, less abundant flashes and glows, small puffs of what looked like luminous blue galactic clouds, and distant orbs that would glow brightly for three seconds and then wink out. I was awe-struck and baffled.

Producing light takes energy—*a lot* of energy. Energy is the cur-rency of life, and it is never spent frivolously. How was this seemingly profligate expenditure possible? Why was it here? And, given that this was obviously the most important thing happening in the ocean, why weren't more people studying it? In those days, it was possible to pick up a textbook on marine biology and find no mention of bioluminescence. And when it *was* written about, it was deemed of little import—akin to fireflies on land, bit players in the grand theatrical production of life. It seemed obvious to me that in this theater, the light emitters weren't bit players; they were the stars, both literally and figuratively.

When Charlie called to say it was time to wrap it up, I couldn't believe it. It felt like no time had passed, and I didn't want to leave. I turned on the lights and looked out into the void, anxious to know who was making all that light and why. There were a few sergestid shrimp and one small jellyfish, but no mermaids. Nothing I saw with the lights on could account for all those tears. *There's nothing there!* I thought. Well, that was certainly a problem, and might explain why more people weren't studying bioluminescence.

As Charlie began hauling up the suit, I turned the lights off again and watched a dazzling meteor shower of luminescence whipping by as my mind raced. What part did light really play in this world? How was it similar to light's role in the terrestrial realm, and how was it different? What kinds of experiments could I devise to answer such questions?

Intermixed with these scientific questions was the beginning of a realization: If you gain access to a portion of the planet as remote and inaccessible as this and it proves filled with glittering treasure, what choice do you have but to come back—again and again and again?

Chapter 5

STRANGE ILLUMINATION

Drifting on a glassy tropical sea at high noon, if you stare over the side of a ship you may observe sunbeams disappearing into the depths. The dancing rays, made visible in the clear water by plankton and particles that scatter light back to your eyes, appear to converge and lead downward into a dark tunnel. The effect is hypnotic, bringing to mind Alice peering through the looking glass. She wanted to see behind the fireplace to know if a fire burned there, as it did on her side of the mirror. The more she looked, the more she wondered what it would be like to live in that other world. In Lewis Carroll's telling, she gets to find out when the glass melts away, "like a bright silvery mist," and she passes through to the other side. The world she finds is exceedingly strange, inhabited by peculiar beings living bizarre lives.

The mirror surface of the ocean divides our planet's living space into two realms: the air-filled one where we reside and the water-filled one that is home to most of the life on Earth. Passing through that air-water interface unveils an equally fantastical world, populated with wondrous life-forms uniquely adapted to live there.

Attempts to understand life and life processes begin with what the Nobel Prize–winning animal behaviorist Niko Tinbergen described as "watching and wondering." That fundamental practice is something we all engage in to varying degrees, but Tinbergen formalized it in the study of animal behavior, becoming one of the founding fathers of what came to be called ethology. Key to his approach was observing animals in their natural environment. He deemed this essential because what animals look like and how they behave is all about adaptations to the places they live.

The immense oceanic realm that covers most of our planet and extends from sunlit surface waters to the seafloor is the least understood environment on Earth. Opportunities for watching and wondering here are rare, yet even a single visit is enough to reveal its most outstanding feature: There's nothing to hide behind!

On land, prey avoid predators by concealing themselves behind trees or bushes or burying themselves in hidey-holes. For midwater prey that inhabit the open water between the surface waters and the seafloor, there are no such options. How do prey evade detection when the only thing between the hunter and the hunted is crystal-clear water?

Imagine you are a fish hanging around in the open ocean at six hundred feet deep. In very clear ocean water, this is roughly where the light intensity drops below 1 percent of surface light levels. Below this depth, there is insufficient light for photosynthesis, but there's still plenty of light by which to see. If a hungry shark swam into view right now, you'd be an easy target. Your best option to avoid being spotted is to swim straight down toward darkness. The trouble with this plan is that you'd be leaving your food source behind. Photosynthesis is the basis of the oceanic buffet, and even if you're not a vegetarian, the critters you eat, be they crustaceans or smaller fish, probably are, which means the best place to find them is around the plants they eat, which is to say the phytoplankton, growing at the surface.

The solution is to use darkness as cover, traveling back up to the surface to eat after the sun has set. Every day, massive numbers of animals in the ocean employ this strategy, vertically migrating toward the surface at sunset and then migrating back down before sunrise. On a ship's sonar, these layers of migrating animals are so dense they look like the seafloor ascending and then descending beneath your ship.

There is probably no starker testament to how profoundly animal forms and functions are shaped by their visual environment than life in the midwater. Understanding animal adaptation in this realm requires visualizing what scientists call the light field. This is defined as all the light streaming in every direction through every point in space, be it in air or water. Defining this is so important to understanding how light behaves underwater that there is a whole scientific discipline dedicated to it, called ocean optics.*

William Beebe described the odd appearance of the underwater light field as "the strange illumination." Those words initially appear in this dramatic description of his first deep dive with Otis Barton in the bathysphere, when they reached the heretofore unimaginable depth of seven hundred feet:

> Ever since the beginnings of human history, when first the Phoenicians dared to sail the open sea, thousands upon thousands of human beings had reached the depth at which we were now suspended, and had passed on to lower levels. But all of these were dead, drowned victims of war, tempest, or other Acts of God. We were the first living men to look out at the strange illumination: And it was stranger than any imagination could have conceived. It was of an indefinable translucent blue quite unlike anything I have ever seen in the upper world, and it excited our optic nerves in a most confusing manner. We kept thinking and calling it brilliant, and again and again I picked up a book to read the type, only to find that

---

*Warning: The field of ocean optics employs mathematical equations that may be offensive to math phobics.

I could not tell the difference between a blank page and a colored plate.

His descriptions of that "strange illumination" were heard again
and again in the reports he sent up the telephone line. So often was
it mentioned that he later confessed that "the repetition of our insis-
tence upon the brilliance, which yet was not brilliance, was almost
absurd."

What Beebe was experiencing and attempting to describe was
how profoundly sunlight is altered when it passes from air into
water. Most apparent is the degree to which different colors are absorbed, causing a radical shift from a multicolored, sun-drenched
palette above water, filled with warm yellows, oranges, and reds, to a
submarine world bathed in a cool vibrant turquoise. But there is also
the impact of scattering, which gives the expression "bathed in light"
special meaning. Were it not for the effects of scattering, the only
visible light would come from directly overhead, and peering into
the distance horizontally would reveal only blackness. Instead, the
water itself seems to be emitting light, a direct consequence of scattering.

With the combination of absorption and scattering, daytime illumination in the midwater assumes a predictable form, best conjured
by imagining yourself in the center of a giant blue beach ball that's
illuminated by the surrounding water. Straight overhead, the light is
brightest. Straight down, it is two hundred times less bright, and
between these two extremes there is a light-to-dark gradient. What is
remarkable about your view is how symmetrical it is. Pirouette in
any direction and it appears *exactly the same*. Incredibly, this sym-
metry does not change, no matter where the sun sits in the sky. Be it
at the peak of its arc, at midday, or traversing the horizon at dawn or
dusk, the only apparent change below two hundred feet deep is in
intensity. This is because the shortest path for light through water
results in the least attenuation, so the brightest sector in your beach

ball is constantly straight up—the point of minimum distance between you and the surface.

Now imagine yourself again as a fish swimming toward darkness in order to hide. How deep do you have to go? Light decreases tenfold for every 250 feet you descend, which sounds like you shouldn't have to swim very far, but eyes—especially eyes adapted to seeing in the deep ocean—are incredibly sensitive, and they can enable the detection of sunlight at depths in excess of three thousand feet, well below what human eyes would deem the edge of darkness. That's a whole lot of swimming to do on a daily basis! For a hatchetfish, it would require swimming seventy-two thousand body lengths each day, the equivalent of eighty-two miles for Olympian Mark Spitz, who even at his peak generally maxed out at twelve miles per day. Clearly, anything that could reduce the length of that swim would represent a tremendous survival advantage. If only there were some way to camouflage yourself so you could blend into the background, then you could stay closer to the surface and not have to make such an exhausting trek between your dinner at the surface and your daytime safe haven in the dark depths.

How can you possibly hide when, from whatever direction you are viewed, you are going to appear as a silhouette against a stark, illuminated background? This is why the hatchetfish has silver sides. Its scales mirror light from the featureless symmetrical orb that surrounds it, reflecting light that is a close match to the light behind it. This even works for predators that are slightly above or below, because the mirrors are roughly vertical even on curved portions of the fish. But what about when seen from directly above? Looking down on the hatchetfish, you will see that its back is darkly pigmented, the better to blend into the darkness beneath it. This kind of camouflage, where an animal is darker on its top side than its underside, is common. Known as countershading, it renders animals such as sharks less conspicuous when seen from above or the side, making it easier for them to approach prey undetected. The word *counter-*

shading is about countering the natural effects of shading. A light shining from above creates a bright back and a deeply shadowed belly. Leonardo da Vinci said, "Shadow is the means by which bodies display their form." Countershading is the means by which bodies hide their form.

But a white belly cannot hide the most conspicuous shadow of all: the silhouette seen from below. To reduce the size of its shadow, many a fish has a thin shape. That body form is not purely a matter of hydrodynamics, as evidenced by the fact that the fastest swimmers in the ocean, such as sailfish, marlin, bluefin tuna, and blue sharks, are round, not slender. However, for a fish to really disappear, it must somehow replace the light that its body absorbs, which is where bioluminescence comes in. This type of bioluminescence camouflage is called counterillumination, and, given how many animals use it, it must be an enormously effective cloaking device.

The thing I find most astounding about counterillumination is how many different ways animals have evolved to achieve the same thing. To create perfect camouflage, the light that these animals emit must exactly match the light field above them. A predator swimming beneath a counterilluminator doesn't see its prey because the fish replaces the sunlight absorbed by its body with a perfect bioluminescent imitation of the missing light. This means that if a cloud passes over the sun and dims the downwelling light, an animal must somehow dim its bioluminescence to match.

Some lanternfish have a photophore just above each eye that the fish can use to compare its bioluminescent output to the overhead light it is trying to match. If the photophores' light outputs look too dark against the background, then it either brightens its bioluminescence or swims deeper until the light matches. But there are many examples of animals that don't have light organs whose output they can see, and it's unclear how they achieve such a perfect match. Perhaps their eyes measure light with much greater accuracy than ours do. Because our eyes can adjust to different light levels, our assessment of brightness is highly dependent on our most recent experience. Just think about how differently you perceive light when you first enter a darkened room compared with how you can see after you've been there for twenty minutes. A light that appears bright after you are fully adapted to the darkness might be invisible after you've been in direct sunlight. This is a problem if you need your eyes to perform as reliable light meters.

Color, too, must match, and many animals have evolved elaborate optical filters that serve to narrow the spectrum of their bioluminescence, producing a purer blue to look like that found in the deep ocean. The hatchetfish's fingernail-like light organs with their magenta lunules do not produce pink bioluminescence, as I have seen suggested in some popular literature, but rather they absorb some of the shorter and longer wavelengths of the fish's natural blue bioluminescence to create a perfect color equivalent. The magenta seen under white light is the result of a combination of red and blue light transmitted through the filters and reflected back to the viewer. That these filters also transmit red light is meaningless, because there is none to transmit down where they live. A filter can't create color; it can only subtract it.

One further way that animals must blend with the background light, besides intensity and color matching, is directionality. There is an artist named Larry Kagan who creates 3D sculptures out of thick metal rods, sort of like rebar. Each sculpture looks like an abstract doodle in space, until you turn on the spotlight that illuminates it. Then there is a moment of epiphany as that light casts shadows on the wall behind the sculpture, revealing the image of a chair in one case, an insect in another, and Che Guevara in another. To appreciate these artworks, you need to embrace shadow as a volume in space, one that is created by a highly directional light source.

Counterilluminators use this concept and have evolved all kinds of tricks to ensure that the light they produce with their photophores has the same directionality—or, more specifically, the same angular distribution in space—as the light they are replacing. Some do it with lenses, and some do it with very clever use of concave mirrors. The hatchetfish does it with fiber optics—mirrored tubes that carry the photons from inside the fish, where they are emitted by light-producing cells called photocytes, through the magenta-colored filters, down to the fish's underside. Where the light emerges, the photophores look like tubes cut on a steep angle that create the fingernail-like appearance of the light organs and help shape the angular distribution of the emitted light.

As any *Star Trek* fan can tell you, cloaking devices are a huge energy draw. This is why, for starships, the moment of cloaking or decloaking provides a window of vulnerability as energy is transferred from shields and weapons to or from the cloak.* The greatest window of vulnerability for counterilluminators is sunset. Since food is concentrated in surface waters, the migration becomes a race to the top. First come, first served. But in order to get to the dinner table without being seen (and devoured yourself), you need to be pumping out enough light to match the light coming from above.

Those that produce the brightest bioluminescence can take the lead in the race to the dinner buffet above while still remaining cloaked. But the more you push the upper limits of that envelope in an attempt to score more food, the more food you have to eat in order to compensate for all the energy you've burned. The daily energy requirement of counterilluminating at 1,300 feet deep in clear ocean water is the equivalent of adding a brisk half-hour walk to your daily routine. At 1,000 feet, it is equivalent to adding a one-hour swim. But at 650 feet, pumping out enough bioluminescence to match the sunlight found at that depth would be equivalent to running one and a half marathons every day—which may not be the

---

* You might be a Trekkie if you know that this was revealed as a weakness of the Romulan Bird of Prey introduced in the first-season episode "Balance of Terror."

best use of your resources. Clearly, animals have a range of environ-mental variables to which they are adapted. Pass too far outside that range and bad things happen—like using up your energy stores be-fore you can replenish them.

AT THE TIME we were making our Wasp dives in 1984, most of what was known about animal adaptations to the midwater light field was based on net sampling. Measuring light at a particular depth, usually by lowering a light meter on a wire from a ship and then dragging a net at that depth, provided some indication of which animals lived in which light zones, but it was a coarse view, with no insight into individual animal behaviors.

Wasp provided an opportunity for "watching and wondering" to a degree never before possible, because not only could I observe the animals in their natural habitat, but I hoped to quantify light's influ-ence on their behaviors using a couple of purpose-built underwater instruments. The first of these was a light meter sensitive enough to measure very dim levels.* The second was something I dubbed the "light wand," which was simply a small blue light in a pressure hous-ing that I had stuck into the end of a three-foot length of PVC pipe. I planned to wield it in front of the Wasp like a light saber, blinking the bulb on and off by using a switch inside the suit, in the hope that it would permit me to talk to the animals, Dr. Dolittle style. Since it seemed obvious that the language of life in the midwater was ex-pressed through light, I wanted to see if I could crack the code. I had visions of either animals flashing in response or predators attacking out of the darkness, and I couldn't wait to try it out.

Everything about my first dive in Wasp was so mind-blowingly

---

* Since no such underwater light meter existed, we had to create one. This was a collaboration that involved engineer Mark Lowenstine and fellow Case Lab gradu-ate student Mike Latz. The light meter was based on a photomultiplier tube built into an underwater housing designed with a gimballed mount that assured that it always pointed straight up.

awesome, I assumed my next would be even more so. Having passed the psych test with ease, I felt confident that I wasn't going to have any kind of claustrophobic meltdown and so I'd be able to focus on my observations. That turned out to be a bad assumption.

Panic is not conducive to sound decision-making, which is why the admonishment "Don't panic!" is so often offered up to those behaving like their hair is on fire. It's great advice, but singularly unhelpful if unaccompanied by how-to instructions. The first time I experienced full-blown panic was in the hospital. Three weeks after the spinal fusion, the discovery of the massive infection in my surgical site required emergency surgery, which was carried out without benefit of general anesthesia because of concerns about retriggering the blood disorder. Before the procedure, a nurse administered a hypo with a standard anti-anxiety cocktail intended to relax me. Instead, it left me feeling like I had been robbed of the only tool I had to fight the pain and fear: my brain. Even worse than the pain was the extreme anxiety I felt every time someone from the surgical team would ask me how my breathing was doing. I was overwhelmed by this rising sense of panic that I couldn't control because my brain was drugged. The surgery lasted an hour and a half. It felt like eons. When the ordeal was finally over, I begged the doctor to promise me that I would never have to go through anything like that again. All he would say was "We'll see."

Two days after the procedure, a nurse showed up in my room, hypo in hand, ready to prep me for another surgery. Her arrival ripped my fingers off the metaphorical cliff face to which I had been clinging, and I was in free fall. There had been no warning from anyone that this was coming. I had no time to prepare mentally for it, and I wigged out.

The fear overwhelmed my capacity to think rationally. I was hysterical, begging my poor mother not to let them take me. The doctor came in to try to calm me down and explained that some more dead tissue needed to be cleaned away but that it would not take nearly as long as the last time. I finally agreed to go quietly if they promised not to give me the drug cocktail, because, as I tried to explain in the calmest voice I could muster, to deal with the panic I needed to be able to think. He was dubious. Anti-anxiety meds must have seemed like just the ticket, given my behavior, but he acquiesced, although the hypo was kept next to my gurney as an unspoken threat.

The procedure was not pleasant, and I had to fight back waves of immense fear and pain, but I got through it. In fact, I got a lot of practice getting through it, because it then needed to be repeated every other day for a month. So here are my how-to instructions for not panicking: *Refocus*. That's it.

It's actually the same advice the White Queen gave Alice to keep from being sad, only instead of calling it refocusing, she called it "considering things."

"Can *you* keep from crying by considering things?" [Alice] asked.

"That's the way it's done," the Queen said with great decision: "nobody can do two things at once, you know. Let's consider your age to begin with—how old are you?"

"I'm seven and a half, exactly."

"You needn't say 'exactually,'" the Queen remarked: "I can believe it without that. Now I'll give *you* something to believe. I'm just one hundred and one, five months and a day."

"I ca'n't believe *that*!" said Alice.

"Ca'n't you?" the Queen said in a pitying tone. "Try again: draw a long breath, and shut your eyes."

Alice laughed. "There's no use trying," she said: "one *ca'n't* believe impossible things."

"I daresay you haven't had much practice," said the Queen. "When I was your age, I always did it for half-an-hour a day. Why, sometimes I've believed as many as six impossible things before breakfast."

Rather than consider "impossible things," my preference is for *intriguing things*, but the Queen was right: The key is *practice*. The ability to refocus your brain when it tries to go off half-cocked in some counterproductive fashion is such a valuable skill, I think it should be taught from an early age.

MY SECOND DIVE in Wasp came two days after the first. It was about an hour before sunset. I was descending through the water column as dusk fell, which meant the edge of darkness was rushing up to meet me. All traces of sunlight were nearly gone at about 350 feet below the surface, where I encountered a distinct layer of krill. This was the vanguard of migrators heading for the surface. It was as if they were raising the curtain on the fireworks display, because from there on down, the luminescence was spectacular.

I had Charlie stop me at 880 feet and then again at 1,400 feet so I could study the fireworks and try out the light wand. There was so much luminescence swirling around the suit, it was difficult to tell if my little light was having any effect. It was like striking a match in the middle of a brushfire: My paltry ember was lost in the blaze. Certainly, no predators were attracted to it. There were a couple of times when I thought something flashed a response, but, given how much other flashing was going on, there was no way to be sure.

As I proceeded down through the water column, I focused on the luminescence, which remained intense until I was about 30 feet off the bottom, at which point it almost disappeared. Several times during my descent, I was distracted by a popping sound. When I described what I was hearing to Charlie over comms, he said, "It's probably the syntactic foam on the outside of the suit."

Syntactic foam is a common solution to the problem of providing flotation under such crushing pressures. Styrofoam floats because of air spaces inside a polystyrene matrix, but when they're put under pressure, those air spaces collapse. Syntactic foam floats because of air spaces inside hollow glass microspheres embedded in an epoxy matrix. Glass is actually quite strong when it's in the form of a sphere, and it stands up well under pressure. It wasn't the glass that was cracking, but the epoxy holding it together, presumably because there were microbubbles in it that were collapsing. At least that was Charlie's theory. He explained that the big form-fitting block of syntactic foam attached to the Wasp's back had nothing to do with the suit's pressure integrity, so I shouldn't worry.

Although the increasingly frequent pops were making me edgy, I trusted Charlie's explanation and was okay with it until I touched bottom at 1,831 feet. That was when Charlie called down to say, "Congratulations! You just broke the world depth record for the Wasp." "What the hell do you mean?" I shot back. "I thought this thing was rated for two thousand feet!" His response, "Yeah, but nobody's ever been," was not reassuring. At that moment, the syntactic foam gave forth the loudest pop yet and I suddenly had a very clear image of how much water was over my head. I might have passed the claustrophobia psych test on my first dive, but today all bets were off.

A column of water one foot by one foot square and 1,831 feet high weighs more than 100,000 pounds. That translates to a staggering amount of pressure. At that depth, the tiniest leak could create a high-pressure jet that would cut through my flesh like a hot knife through butter. It had taken eighty minutes to reach this depth. Even if they pulled me up at full speed, it would take at least thirty minutes to get to the surface. I felt panic start to gain a stranglehold, and all I could think was *GET ME OUT OF HERE!* Just as I was about to give full-throated voice to that thought, a jellyfish caught my attention. It had an iridescent thimble-shaped bell and long, flowing tentacles, and it was swimming fast by pulsing its umbrella. Suddenly it dropped its tentacles in a swirling, tangled mass as it disappeared into the darkness.

Focusing on that jellyfish helped pull me out of my vortex of panic and back into a far more palatable reality. As I had learned to do in the hospital, I controlled the fear by refocusing, putting all my attention on the exquisite beauty encompassed in that fragile being, apparently perturbed by my presence but blissfully unaware of the weight on its back. Buddhists may claim that the best way to soothe the monkey mind is by asking yourself, *What's the worst thing that can happen?* But I have found that this is not the best strategy while on the bottom of the ocean. It's far better to simply focus on something else, and if that something is simultaneously magnificent and mysterious, so much the better.*

THROUGHOUT EVERY ONE of the dives I made in the Wasp, there was no shortage of mysteries to draw my attention. Besides the bioluminescence, which continued to baffle and mesmerize, there was also the daily cavalcade of migrators to ponder. I had a series of dives planned using the light meter to establish the light levels where different animals were found at midday compared with those at dawn and dusk.

There were two populations of shrimp I was especially interested in observing. These were the shrimp on skis that I had seen during my first dive and the krill I had encountered on subsequent dives. Both are bioluminescent counterilluminators, but their light organs are radically different in both form and light-producing capacity. Krill generally have ten photophores—one under each eye and eight on the underside of the body—all capable of very bright light emissions. The body photophores are exquisite optical constructions almost like little eyes, except that instead of collecting light, they emit it. In place of a retina, there is a group of light-producing cells called the lantern, beneath which a lens and an iris help to collimate the light before it shines downward. Around the back of the photophore, above the lantern, a reflective layer of cells helps focus the light downward, and around that, a layer of red pigment further assures

---

* I held the depth record in Wasp for a grand total of two days, until Robi found a deeper place to dive and usurped the title. He was welcome to it.

that no light escapes upward. Intriguingly, the photophores on the eyes lack the lenses found in the body organs, perhaps because they are used as flashlights. By contrast, the shrimp on skis, better known as sergestids, have far less elaborate light organs, capable of much dimmer light emissions. These light organs are actually formed from modified liver (hepatopancreas) cells and, although simpler in design than krill photophores, they, too, include a lens to assure that the direction of the emitted bioluminescence matches that of down-welling sunlight.

Of course, all this careful matching to the background light is for naught if the animal tilts its body relative to the downwelling light. This is potentially a problem for vertical migrators like these shrimp, which must incline their bodies up or down to change depth; to compensate, their light organs counter-rotate in order to always maintain their vertical orientation. Krill can rotate them as much as 180 degrees, which allows them to swim nearly straight up or straight down without risking any misalignment of their biolumi-nescence with the background light they need to match. Sergestids, too, can rotate their light organs, in their case as much as 140 de-grees. Hatchetfish, on the other hand, have an equally amazing ad-aptation: They can swim diagonally upward or downward without tilting their bodies. It is this capability that accounts for its ungainly shape, a necessary adaptation that allows it to present a streamlined profile for both horizontal and diagonal swimming.

It was Richard Harbison, one of my fellow Wasp pilots, who dis-covered the hatchetfish trick a year before our Wasp dives. While descending in a submersible, he noticed several hatchetfish swim-ming in a straight horizontal line across his field of view. He watched them for a little while before it dawned on him that this made no sense, because the sub was dropping at a rate of almost two feet per second. The only way the fish could appear to be swimming hori-zontally, he realized, was if they were actually swimming diagonally at a pretty impressive clip. The necessity of living in the ocean's strange illumination is the mother of some mighty fantastic inventions.

I got to see some of this fantasticness firsthand during my dives with the light meter. The first of these took place at high noon so that I could make careful notes on where different animals were found relative to different light levels. There were clear delineations. A smaller species of krill (*Euphausia pacifica*) formed a layer between about 550 and 650 feet. At about 750 feet, another layer of krill, a larger species (*Nematoscelis difficilis*), occupied a light level twenty-five times dimmer. And below 1,000 feet, the sergestid shrimp resided in a murkiness that was two hundred times dimmer still, right at the limit of my eyes' and the light meter's ability to detect any light coming from the surface.* To see it, I had to look straight up, and all I could detect was a hint of gray surrounded by the blackest of black. That this is enough light to reveal a silhouette and thus warrant expending energy on counterillumination is a testament to the sensitivity of the eyes of deep-sea predators. To achieve such sensitivity, they often sacrifice acuity, with adaptations like neurally linking neighboring photoreceptors, which helps explain why the belly lights found on so many fish, squid, and shrimp, which appear as distinct points of light to our eyes, might actually blur together, forming a diffuse glow that blends perfectly with the background as seen by the eyes of deep-sea predators.

I also made dives at dawn and dusk, hanging at a depth of 500 feet while watching the commuters swim by. I thought I would see a well-ordered migration with a staggered order of appearance as animals adhered to their preferred light levels, but it was more chaotic than that. Some stuck with their light zones and some didn't. At sunset, the small and large species of krill were mixed together, with

---

* In coastal waters, particles and dissolved substances in the water increase absorption and scattering, so light doesn't penetrate as far as it does in clear ocean waters.

the large krill taking the lead, hightailing it for the surface and out-swimming their smaller brethren. The krill reached the depth at which I was hanging before sunset and continued to pass by for more than an hour afterwards as the light level dropped more than three-hundred-fold. Then the first sergestid shrimp came into view, and they reached their peak concentration an hour and a half after sunset, apparently following the same very dim light levels they occupied during the day. Gooseberry comb jellies and corncob sipho-nophore chains (*Nanomia bijuga*) were also racing for the surface, with the comb jellies in the lead. And mixed in with the sergestid shrimp were fish and squid. Like rush hour in most of the world's big cities, it lasted longer than an hour, but it felt much better managed.

To be a firsthand observer of this massive daily trek was an extraordinary privilege. On my first sunset dive, as I held stationary at 500 feet, I got to watch the edge of darkness well up from the depths and pass me by on its inexorable ascent. What peculiar and exhausting lives these creatures lead, spending their existence as commuters, racing toward the surface at sunset, seven days a week, and descending back into the depths at dawn. As a result, they live forever in darkness, which helps explain why so many of them are bioluminescent. The best way to cope with perpetual night is to make your own light. That might sound like something you read in a fortune cookie, but it's an evolutionary concept that considerably predates cookies of any kind.

As I dangled there, I kept my lights off, turning them on only briefly to survey the scene and count the commuters visible within my field of view. When the lights were off, I was watching the bioluminescence. Up until sunset, there was none, either because the downwelling sunlight made it impossible to see or because it wasn't there. But as the sun set and the light faded, I began to notice a very few brief flashes appearing, initially at fewer than three per minute and then gradually increasing to a crescendo of both short and long flashes and a combination of point sources and short, fragile chains of mermaid's tears. The light show reached its peak an hour after sunset, when the flashes became too frequent to count; I had to estimate their density by the spaces between them, which I guessed as two inches between the point sources and anywhere from two to six inches between the mermaid's tears.

The more I studied the bioluminescence, the more I began to be able to guess who was who. I suspected that some of the brighter point sources were krill that were stimulated to flash by the undulations of the Wasp. Some of the point sources looked like a background of sparkling pixie dust, and I guessed these might be dinoflagellates. But there was so much I was seeing about which I had no clue. One type of flash in particular intrigued me. It was a bright, slow-on, slow-off flash that lasted about five seconds and was so far in the distance I was sure it wasn't being mechanically stimulated by the suit. I routinely saw it associated with the sergestid shrimp layer, both deep during the day and migrating with it at twilight.

When I was first reviewing my dive transcripts and saw the entry "bright flash in the distance," I had to sit and think for a while: How did I know it was "in the distance"? I tried to re-visualize it, and when I did, I remembered the halo that surrounded it. It was the scattering of the light that gave its remoteness away; the bigger the halo, the more scattering, the greater the gap. This must be an important clue for animals having to judge distance in the dark. If you see a flash and swim toward it, you need to have some notion of how far to swim.

If we want to understand life on our planet, we need more time for watching and wondering within its blue heart. Light evidently plays a critical role here, but often in ways we do not yet fully comprehend. As the Earth spins on its axis and the intensity of sunlight impinging on the surface of the ocean waxes and wanes, the push and pull of twilight's threshold on myriad creatures in the depths is relentless. An overcast sky can cause multitudes to relocate to shallower depths or dim their bioluminescent belly lights. If sunlight exerts such sway, then what of bioluminescence itself? Living light dominates the submarine light field at depths below the penetration of sunlight, and at night in surface waters, but it is poorly characterized compared with the solar light field. I desperately wanted to know what the true nature of that biological light field was when there was no big mechanical suit down there stirring things up. These were clearly solvable problems, I realized; we just needed to find new ways of observing.

Chapter 6

A BIOLUMINESCENT MINEFIELD

The day of the flood dawned calm and cloudless, with no hint of impending doom in the air—if you disregarded my journal entry scribbled the night before: "Lousy sleep. Keep having dreams of entrapment and drowning."

A year had passed since the Wasp expedition. Although the original plan was for another mission using Wasp, the suit had proven less than ideal for our purposes. The biggest hindrance was the tether. That connection to the ship made for a wild ride when the sea was rough, transferring all the wave action at the surface down the cable to the suit's occupant. During the course of a dive, we could reduce this to some degree by adjusting the ballast. Putting slack in the tether provided a certain amount of cushion, but I never felt fully decoupled from the tea-bag motion, so I couldn't be sure how much of the bioluminescence I was seeing was stimulated by the suit. Also, during descents and ascents, when there was full tension on the cable, the experience felt akin to being inside a martini shaker.

All the headaches caused by the tether, both literal and figurative, led Robi to look for an alternative. He settled on the untethered single-person submersible Deep Rover, which was the most recent brainchild of the same inventor who'd developed the Wasp, Graham Hawkes. Deep Rover appeared to have all the advantages of Wasp for providing direct access to the midwater, and none of the drawbacks.

In Wasp you are always standing and, because the metal body sucks the heat out of you, frequently cold. Deep Rover, meanwhile, is more like an underwater helicopter, where you sit in a comfortable pilot's seat in the center of a nearly invisible five-foot-diameter acrylic sphere with five-inch-thick walls that insulate you from the cold. Even better for our purposes, it did not rely on a tether, which meant that there was no martini-shaker effect. This made Deep Rover potentially the perfect platform for answering one of the biggest questions related to bioluminescence in the ocean: How much occurs when we're not down there stirring things up?

Virtually all the natural light that illuminates life on Earth originates from two sources: the sun and bioluminescence. Scientists know a lot about how profoundly sunlight impacts animal adaptations and behaviors in the ocean above the edge of darkness. Below the edge of darkness, the presence of so much bioluminescence and so many animals with eyes suggests that living light is equally impactful. But in what ways? Although the field of ocean optics had gone a long way to describing the solar light field, there was very little understanding of the bioluminescence light field and how it affects animal behaviors.

The first time sensitive light detectors were lowered into the ocean, in the 1950s, scientists were amazed by the amount of light they recorded. The light meters were designed to measure the penetration of sunlight underwater, but once they dropped below a thousand feet, they started recording flashes. At first, the investigators wondered if there was something wrong with their detectors, but then they realized it must be bioluminescence. These were bright flashes, and there were lots of them. At two thousand feet, flash intensities were a thousand times greater than the intensity of sunlight at that depth, and flash frequencies exceeded one hundred per minute. It called to mind a Disney light parade with an extravaganza of illuminated floats and fireworks spectacles. They wondered, *What the heck is going on down there?*

Since the greatest amount of flashing was observed in the vertically migrating layers seen on sonar, one suggestion was that all this light output might be connected to an increase in the animals' metabolism needed to make such long migrations. Another idea was that maybe the flashing served to help coordinate rush hour traffic by keeping the commuters in sight of one another. Although these notions sound far-fetched now, they reflect how little was understood at that time about the functions of bioluminescence.

A number of papers were published detailing flash frequencies at different depths and different times of day, until eventually it was noticed that the frequency of flashing correlated with the sea state: Rough seas generated more flashes than calm seas. The researchers deduced that their instruments must be bumping into light emitters, causing them to flash. So the question became: What is the true background level of bioluminescence? It was surprisingly tough to answer. Decoupling any detector on a cable from the motion of the ship on the surface is a major challenge. And it doesn't help to anchor the detector to the bottom, because currents flowing around it also mechanically stimulate bioluminescence.

Being able to determine the levels of spontaneous bioluminescence was a big deal for two reasons. The first—the one *I* cared about—was that it spoke directly to understanding the nature of the visual environment in the biggest living space on the planet. If I was ever going to be able to comprehend what life is like for the animals inhabiting this space, I really needed to know what the visual scape looked like in its undisturbed state.

The other reason it was important had to do with military concerns. The U.S. Navy was looking into using lasers as an acoustically quiet means of underwater communication for submarines. They wanted to know what kind of signal-to-noise ratios they might expect. If there was a lot of spontaneous bioluminescence, that equated to a high level of optical noise, and potentially muddled communications.

It seemed like Deep Rover might finally provide the means to answer this question. According to Graham Hawkes, his little sub had such good ballast control that it could essentially become one with the surrounding water.

In order for a craft to be neutrally buoyant, the force of gravity pulling down must be exactly offset by an equal but opposite force of buoyancy pulling up. The result is that you neither sink nor float. Some fish control their buoyancy by means of a swim bladder— a gas-filled sack that they can inflate when they want to ascend, deflate to sink, or adjust to perfectly offset the downward pull of gravity when they want to hang out at a particular depth. Deep Rover has something similar in the form of a soft ballast system that feeds compressed air or water into a tank to make the sub either lighter or heavier. I hoped to be able to trim the sub so perfectly that I could eliminate any mechanical stimulation of bioluminescence and then sit quietly, watching a world that didn't know I was there and counting spontaneous bioluminescent events.

We had the same team of scientists as the year before—Bruce Robison, José Torres, Larry Madin, Richard Harbison, and me—but because Larry had to leave early and Rich had to arrive late, Robi, José, and I were going to get the lion's share of the diving. Deep Rover was a lot bigger than Wasp, which meant we couldn't train in a tank the way we had the previous year. Instead we were handed a manual to memorize and given some classroom training in emergency procedures, and then it was time to dive.

For this expedition, we had moved north from Santa Barbara to Monterey Bay, home to one of the world's most spectacular submarine canyons—comparable to the Grand Canyon in depth, but with steep escarpments and multitier plateaus studded with all manner of marine life. The canyon also serves to funnel deep-sea animals from offshore up the chasm, potentially providing a greater concentration of the midwater animals that we wanted to observe.

My first dive was a training dive near the head of the canyon in water that was only sixty feet deep. They dropped me in with the ship's crane and then held me on the hook while we did pre-dive checks of thrusters, electronics, and comms. By far the biggest drawback of not having the direct electrical link that the tether made possible was a significant degradation of communications. At the surface, we could communicate with VHF walkie-talkies, but as soon as we submerged, we had to switch to through-water acoustic communications, which use the water to carry the signal. Some of the time, this was almost as good as talking on a cellphone, except you had to push to talk, which meant that only one person could speak at a time and you had to remember to say "over" when you were done. But a lot of the time it was a noisy link, subject to dropouts and ocean noise, including the chatter of dolphins. It worked best when restricted to succinct reports of depth, cabin pressure, and oxygen level. No trading bad jokes, as was our wont with Wasp.

Once all was deemed A-OK, they set me free from the crane, with instructions to motor along the surface toward one of two buoys that were part of our training course. At the first buoy, I was told to submerge and then resurface, at which point a scuba diver removed a small fifteen-pound lead weight from the sub. When I tried to submerge again, I couldn't. That meant they had done their buoyancy calculations correctly. The diver replaced the weight and I was instructed to submerge and proceed on a compass heading to buoy number two. I came up very close to it, and then after submerging again I got permission to "go play."

Flying Deep Rover was like the best video game *ever*. It was incredibly easy and intuitive. All the controls were in the seat base or the armrests, so my view was unobstructed. Handgrips at the end of each armrest controlled two multifunction manipulators that responded to the lightest touch. To activate the thrusters, I only needed to slide the armrests forward or backward. I found that if I slid one forward and one back, I could make the sub spin like a top, and the manipulators were so dexterous I could pick anything off the bottom with great delicacy.

The view was panoramic and offered up plenty of eye candy on which to feast: fluffy white-plumed anemones, pink and orange sea stars, bottom-dwelling flatfish, and much more. In just this one short dive, I saw five small octopuses, a diving bird (a grebe) that I observed with amazement at a depth of forty feet, and a sea lion that swam by with such speed that it put the sub's zippy three knots to shame.

My second dive, to 120 feet, was far less thrilling, because the visibility was almost zero and I had to head up early because of bad weather. On my third dive, though, things got *really* interesting. The first two had been considered training dives, and this was my first real science dive, to a depth of a thousand feet for a total of four hours. I went down at three in the morning. Inside the sphere I had a supersensitive video camera* that I hoped would be capable of recording the bioluminescence outside the sub. Up until this point, the only people who had ever seen bioluminescence in the deep ocean had been those lucky few who had been able to dive in a submersible and had bothered to turn out the lights. I badly wanted some way of recording it so I wouldn't have to depend on my visual memory of fleeting flashes, and also so I could share something I considered one of the most beautiful natural phenomena on the planet with people completely unaware of its existence.

As soon as they let me off the crane hook, I cruised away from the ship and then flooded my ballast tanks and began to sink through the inky black waters. I kept my floodlights on as I descended so I could observe the animal life. Almost immediately, I entered a layer of squid and what appeared to be red octopuses. The octopuses reminded me of the red crabs I had seen from Wasp—another creature that one usually associates with the bottom. These were the

---

*The intensified silicon intensified target (ISIT) video camera used two stages of intensification to achieve a sensitivity nearly comparable to that of fully dark-adapted human eyes, although it took black-and-white images and had lower resolution.

ruby octopus (*Octopus rubescens*), which spends a longer portion of its young life drifting about in the water column and growing to a larger size than most other octopus species, before settling down to a more sedentary existence on the bottom.

Descending farther, I encountered many of the same creatures I had seen from Wasp: krill, shrimp, and jellies. There was also a heavy concentration of the white, flocculent marine snow, an indication of the rich planktonic life at the surface. I tried flicking off the lights briefly and immediately saw streaks and swirls of bioluminescence skidding up and over the acrylic sphere. I was torn between wanting to revel in the light show and needing to see who lived here, but, with some difficulty, I chose data over aesthetics and left the lights on.

At seven hundred feet, I slowed my descent by releasing compressed air into the ballast tank. It proved easy to trim the sub out to neutral buoyancy. The depth gauge readout was too coarse to provide the precise feedback I needed but by observing my position relative to the marine snow in the water,* I could readily tell whether I was too heavy or too light. As soon as I was neither, I turned out the lights, prepared to count flashes per minute with the aid of my digital watch and its push-button microlight.

I waited and watched, expectantly peering in different directions, straining to see the tiniest flicker. Seemingly interminable minutes ticked by. Nothing. In front of me was just an enormous, absolute blackness as complete as in the deepest, darkest cave. In our world, where the night is awash in light from all manner of fluorescent bulbs, streetlights, car headlights and taillights, neon signs, cellphone screens, illuminated digital clocks, and 3D Terminator nightlights,† not to mention natural light from the moon and stars,

---

*Although scientists describe marine snow as "raining down" from surface waters it's a slow-motion rain, traveling only three to six inches per minute.

†Available for $99, because who wouldn't want an overpriced nightlight guaranteed to scare the crap out of you at two A.M.?

such all-encompassing darkness is a rarity that few have experienced and many, I think, would find unsettling.

I didn't find it scary, but it was disconcerting, because it was so at odds with what I had expected. After a few minutes I tried tapping the thrusters, and immediately geysers of sparkling specks and fragments of living light erupted out of the propellers. Smatterings of flashes and small puffs of lucent blue clouds blossomed around the sphere, creating a vibrant halo. It was startlingly bright, because my eyes were so thoroughly dark adapted.

As the flashes faded and I was once again swathed in darkness, I tried to think about the significance of what I was seeing. There were myriad luminescent sources; they were all around me. But there was no light unless I triggered it. And all it took was the tiniest movement. I was sitting in the middle of a bioluminescent minefield!

Somehow, animals must negotiate their way through a world in which any movement can trigger a flash that will reveal their presence to the eyes of hungry predators. Imagine you are trapped in a pitch-black Superdome. There is food in the form of yummy apples dangling from strings, if only you can find them before you starve to death. Trouble is, you're sharing this space with a hungry black panther. You can't see him, and, in the darkness, he can't see you. You're safe for now, but for how long? You need to find those apples, but when you try to move, you discover tiny LEDs, also dangling from strings throughout the arena, that light up on contact. Eventually, when your hunger becomes overpowering, your search for food will accidentally trigger a flash and, with a surge of adrenaline, you will realize that the panther's head just snapped around, fixing your exact location.

How can you possibly survive in such a world? One line of defense might be to spew bioluminescence of your own into the face of your attacker, causing a distraction and temporary blindness while you beat a hasty retreat. The fact that many animals use this trick suggests that it is, in fact, effective. Some copepods can release clouds of bioluminescence out of glands on their tails, and there are shrimp that can spew intense streams of liquid light from their mouths, like fire-breathing dragons.* There are squid, like the aptly named fire shooter, that discharge photon torpedoes of blinding blue brilliance. There are even a few fish that can eject sparkling dust storms of light out of a tube on each shoulder, giving them their common name, shining tubeshoulder. These defensive strategies make perfect sense when you imagine trying to play hide-and-seek in a minefield where the slightest disturbance triggers an explosion of fireworks.

As I hung there in the darkness, observing nothing, I imagined predators straining, as I was, to see any hint of a flash. How patient must they be? How patient would *I* need to be to observe any spontaneous luminescence? I had no doubt that it occurred, but apparently it happened on much longer time scales than I had imagined.

Also, I began to wonder, how unobtrusive was I, *really*, sitting here in the darkness? I had brazenly blundered into this space, with spotlights blazing and thrusters shattering the peace. Even with my thrusters off, I wondered what kind of impact the whirring of my scrubber fans might have and what kind of electrical fields might surround the sub. And, although I had made every effort to cover all the indicator lights on the control panel with a black cloth, now that I was dark adapted, I could detect tiny hints of light peeking out here and there. To creatures that have evolved to survive in such a world, I was probably about as unobtrusive as an elephant tiptoeing through a picnic.

I repeated the same observations of complete and total blackness at eight different depths. It was simultaneously intellectually intriguing and incredibly boring. Given the significant cost of this expedition, it was hard not to envision the dollar signs rolling over in

---

*Or in some cases, like that first shrimp (*Gnathophausia ingens*) that I pulled from the trawl bucket, out of nozzles on either side of the mouth.

my head every minute that I sat there, waiting and watching. This certainly did not seem like an effective use of my all-too-limited dive time. Also, if I was going to report zero spontaneous biolumines- cence, I realized that an obvious response might be that there were just no bioluminescent animals in the vicinity. This wasn't true, as I saw when I jabbed the thrusters. Still, just to say there was "lots of flashing whenever I moved" wasn't going to pass muster for a sci- ence publication. I needed to figure out how to quantify the nature of the minefield.

The intensified camera might provide a solution if it could re- cord the bioluminescence. I turned it on, aimed it out in front of the sub, and then backed up, creating a swirling light storm. The cam- era registered bright streaks of light, but they were out of focus and too chaotic to count. To record the bioluminescence, I needed to have the lens wide open, which meant the depth of focus was very limited. I had to figure out a more controlled method of stimula- tion, one that would occur in the constrained plane of focus of the camera.

Back on dry land, I conferred with Robi and his technician Kim Reisenbichler about how to meet this challenge. Robi had brought along a one-meter transect hoop—sort of a metal hula hoop—that he had been planning to secure in front of the sub in order to count the number of jellies passing through it while he motored along. If we knew the forward speed of the sub and the area of the hoop, we'd be able to estimate the number of jellies per cubic meter within a horizontal tube of ocean at this particular location. We reasoned that the same could be done for bioluminescence by stretching some fine-mesh netting across the hoop. If I focused my camera on that screen, then anything that bumped into it and was stimulated to flash would do so in focus. Kim found some five-millimeter (0.2- inch) mesh fishing net, secured it to the hoop with tie wraps, and mounted the hoop on the front of the sub.

The first time I got to try out the screen was a week later, on my sixth dive. It was a clear night under a half-moon, and the seas were calm. I kept my lights on during my descent to observe the distribution of animals in the water column. In the top two hundred feet, there were krill, small fish, and jellies. Just below them were sergestids—the shrimp on skis—and lots of the ruby octopus. I watched intently as one octopus came into contact with the sub's dome and then squirted a stream of reddish-brown ink as it swam away. This was a different kind of tactic for distracting a visual predator, one that would seem to have no purpose in the dark but made perfect sense if it allowed the octopus to cover its tracks through a bioluminescent minefield.

As I proceeded into deeper waters, below eight hundred feet, there were large numbers of fish, especially hake, a splendid silver fish related to cod and haddock. About the size of a small scuba tank, but with a streamlined fusiform shape sporting triangular fins, large eyes, and outsized mouths, they acted curious, cruising in close to the sub with no apparent apprehension. Just to be contrary, some animals seemed attracted to my lights instead of repelled by them.

I continued down, descending through the fish layer until I was 100 feet above the bottom in that spot, at 1,840 feet, where I trimmed the sub to neutral. This time I had brought some black electrical tape with me, and I spent some time covering indicator lights so as to make the sub pitch-black when I turned out the floodlights. I focused the intensified camera on the netting stretched across the transect hoop, shut off all the lights, and turned on the video recorder. Still no luminescence. I recorded a minute and a half of blackness until I couldn't stand to wait any longer, and then activated the forward thrusters. Instantly, bioluminescence was stimulated as it came into contact with the screen. There were transient discrete flashes that passed through the net like blue sparks, small secretions that looked like puffs of neon blue smoke, and fragile amorphous mucous blobs that fragmented and glowed. There were a few creatures larger than the mesh size that were too slow to outswim it.

Bioluminescence in front of the submersible: like the Fourth of July! E. WIDDER

Different bioluminescent displays, clockwise from top right: squid (*Abralia veranyi*), krill (*Meganyctiphanes norvegica*), dragonfish (*Melanostomias bartonbeani*), jellyfish (*Periphylla periphylla*), and worm (*Tomopteris* sp.). E. WIDDER

Edie and David on their wedding day. JAMES MOLLOY

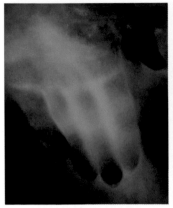

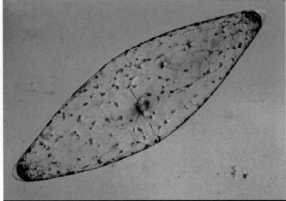

Dinoflagellates poured over the author's hand. TOM SMOYER

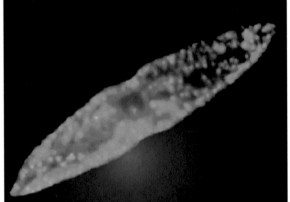

The dinoflagellate *Pyrocystis fusiformis,* as seen with the lights on (top) and as photographed by the light of its own bioluminescence (bottom). E. WIDDER

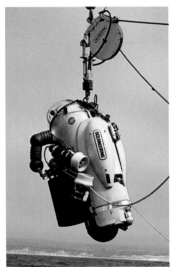

The deep-sea diving suit Wasp, designed for diving on oil rigs to depths of 2,000 feet. E. WIDDER

Ready to dive, bundled up against the cold that rapidly penetrates the metal suit.

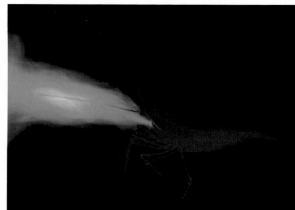

The deep-sea shrimp *Heterocarpus ensifer* spews bioluminescence from its mouth like a fire-breathing dragon. SÖNKE JOHNSEN

Unidentified anglerfish.
E. WIDDER

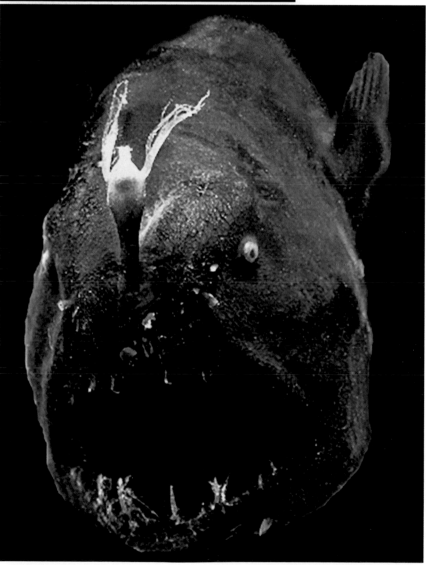

Animals camouflage their silhouettes by producing light from their bellies that matches the intensity of downwelling sunlight. Krill like *Meganyctiphanes norvegica* (left) produce light from ten photophores. Hatchetfish like *Argyropelecus* (center) and lanternfish like the myctophid (right) also use distinctive patterns of photophores on their bellies for counterillumination. E. WIDDER

Left to right: José Torres, Edie Widder, and Bruce "Robi" Robison with the Deep Rover in Monterey Canyon in 1984. SEA STUDIOS

Bioluminescence hitting the SPLAT screen in the Gulf of Maine, showing the different character of the minefield at different depths. Top: a thin layer of bioluminescent copepods (*Metridia lucens*) at 53 feet; field of view is 3.28 feet across.

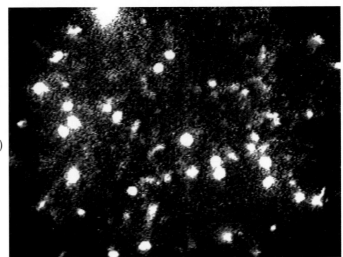

Center: predominantly dinoflagellates (*Protoperidinium depressum*) at 200 feet.

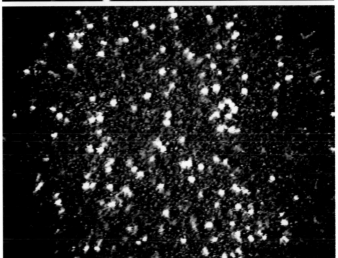

Bottom: a jelly-dominated region at 816 feet where a comb jelly (*Euplokamis* sp.) was responsible for brilliant explosions of light, producing large clouds of bioluminescent particles that passed through the SPLAT screen. E. WIDDER

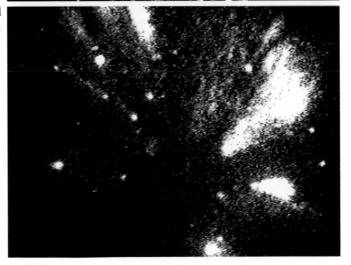

HIDEX-BP
High Intake, Defined Excitation BathyPhotometer

Pressurized Electronics Bottle
Optical Window

Fiber Optics

Pump

Stimulation Grid

First field tests of the HIDEX-BP.
Left to right: Jim Case, Steve
"Bernie" Bernstein, Edie Widder,
Mike Latz, and Dan Ondercin.
COURTESY OF THE AUTHOR

Graphic showing the interior
design of the HIDEX-BP. Inset
(bottom right) shows
bioluminescence stimulated by
the HIDEX-BP stimulation grid,
photographed through a
transparent full-scale model of the
HIDEX-BP. E. WIDDER

Author (left), preparing to pull a gulper eel from the detritus sampler.
COURTESY OF THE AUTHOR

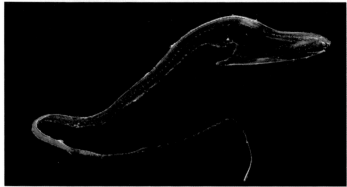

The gulper eel (*Eurypharynx pelecanoides*). E. WIDDER

The Johnson-Sea-Link submersible carries four people: a pilot and passenger in the observation sphere, and a sub-crew member and passenger in the dive chamber. HBOI

Edie in the dive chamber running the spectrometer. COURTESY OF THE AUTHOR

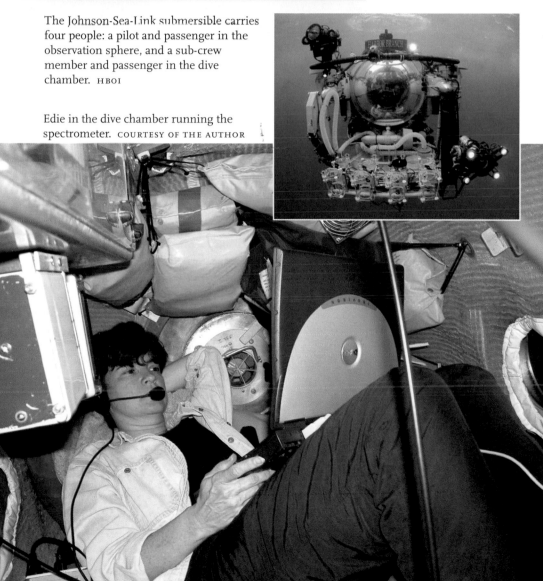

The glowing sucker octopus (*Stauroteuthis syrtensis*). E. WIDDER

Al Giddings (right) describes the workings of the Johnson-Sea-Link submersible to Fidel Castro (left), with the help of Fidel's translator, who stands between them.
E. WIDDER

Instead, they stuck to the screen and glowed, in some cases revealing their shapes, such as the long gelatinous chain of a siphonophore or the clear circular outline of a medusa.

On the camera, the scene looked like nighttime news footage of anti-aircraft fire, with streams of tracer rounds and bomb flashes exploding all over the screen. This was what my college professor meant when he described the wonder and exhilaration of discovery. Something no one had ever seen before. Finally, I was able to quantify a phenomenon never before observed in such multidimensional detail—I was actually able to say how many bioluminescent sources there were per cubic meter. But this wasn't just about numbers; it was also breathtakingly beautiful.

I ran horizontal transects, recording the stimulated bioluminescence at different depths throughout the water column. Everywhere I looked—from surface to bottom—there was a bioluminescent minefield waiting to explode. But the nature of the minefield wasn't always the same. Densities varied, as did the intensities and types of the light emitters. I wondered if animals had to adjust their swimming patterns based on the kind of bioluminescent obstacle course they needed to negotiate. Would an animal swim differently amid a dense array of dim dinoflagellates having a lower threshold for mechanical stimulation than it would through less abundant but much brighter copepods with a higher stimulus threshold? Or might they actively avoid certain kinds of minefields entirely? Light is the most important environmental variable in so many of our planet's ecosystems. Down here, below the penetration of sunlight, did that supremacy still hold true?

With subsequent dives, I continued to run transects at various depths, but I also began trying to figure out who was making which displays. The classic comment of those observing bioluminescence from submersibles that "when you turn on the lights there is nothing there," was as true as ever. Even when there was a large source stuck to the screen and glowing, if I tried to illuminate it with the floodlights, I couldn't see anything. This was because almost all of these larger sources were transparent jellies. At first, when I observed such a display, I tried flipping on the lights and then backing up the sub to try to dislodge it from the screen so I could see it and maybe capture it with the suction sampler, but when that proved ineffectual, I adopted a different strategy—one that I think would make a good arcade game to train future submersible pilots.

Since jellies were prevalent throughout the water column and almost universally bioluminescent, I decided to focus on them. To determine what kind of display a particular jelly produced, I would steer the sub until I had the jelly lined up between the acrylic sphere and the screen. Then came the tricky bit. I would try to simultaneously turn out the lights and back up the sub, while attempting to bull's-eye the jelly in the middle of the screen, where I had zoomed in the camera. If I got a decent recording, I would then maneuver the sub to get the jelly off the screen and in front of the suction sampler, to collect it for species identification and further study at the surface. It took some practice, but eventually I got pretty good at it and was rewarded with some unexpectedly elaborate displays.

Some firm-bodied round jellies emitted a perfect necklace of light when they hit the screen. By contrast, an exquisite crystalline jelly that looked like the top half of a hot cross bun fringed by hundreds of threadlike tentacles did something very unexpected. Whenever I lined it up to strike the screen, it would react to the water currents by contracting its margin into sharp folds, so when it made contact, it was recognizable by its un-jelly-like shape, sometimes forming a near-perfect square of light. Even more remarkable than some of the odd shapes were the incredibly elaborate flash patterns. There were comb jellies generating bands of light that propagated along their comb rows, creating a delicate tracing of a figure eight. More stunning and unexpected were the siphonophore chains that produced dissimilar displays from different parts of their bodies. The common siphonophore (*Nanomia bijuga*) gave a steady glow from its corncob of swimming bells, while the light emitted from its ropy, tentacled lower half scintillated. Another siphonophore, known as *Apolemia*, delivered brightly glowing sources from its lower half and propagated bands of light that danced along the inner stem to which its swimming bells were attached. Many of these creatures are so fragile that they had never been captured intact and no one had ever observed their bioluminescent displays before. I was witnessing a mysterious, sparkling pageant performed just for me.

The elaborateness of the displays was perplexing, especially when you consider that these creatures lack image-forming eyes. So at whose eyes were these showy pyrotechnics directed? I decided to resurrect the light wand and try again. It couldn't imitate displays as elaborate as those I was recording, but I wondered if I might be able to see different responses to a glowing source than what I got with a flashing one. I wanted to test the theory* that glowing sources serve as attractants, while flashing sources act as repellents. We secured the light wand to Deep Rover for my ninth dive, which was scheduled to launch midmorning. This dive did not proceed as planned.

DIVING SUBMERSIBLES, OF course, does not come without risk. The four key areas of concern are the submersible itself, the launch-and-recovery system, the dive site, and the sub crew. We were pushing the envelope on all four fronts. Deep Rover was a prototype without much of a track record. The launch-and-recovery system was a kludge that depended on dangling 3.6 tons of sub from the crane hook while trying to keep it from turning into a wrecking ball by using just rope tag lines and muscle. The dive site was unexplored and the depth profile erratic, which meant it was difficult to keep from drifting outside our safe operating range. Exceeding the range is frowned upon, because if the ballast control malfunctioned,

---

*An idea put forth by scientist Jim Morin.

the sub could sink below what was ominously referred to as its "crush depth." And finally, there was the sub crew.

Over many years of diving in submersibles, I have come to recognize that the sub crew is the secret sauce—the absolutely essential part of any recipe for a safe diving experience. The key ingredient for that sauce is the submersible operations coordinator, sometimes called the SOC—and sometimes "Big Daddy." Charlie had been the SOC for my Wasp dives in 1984. For the first Wasp expedition, in 1982, where I wasn't a pilot, the SOC was Steve Etchemendy, a.k.a. Etch. Both Charlie and Etch had many years of experience as divers and supervisors and had all the attributes that are crucial to the job description. These include strong team leadership, great problem-solving skills, especially under stress, a good sense of humor, extreme attention to detail, and a high degree of what's known as situational awareness, preferably including eyes in the back of the head.

Peter, the SOC that Can-Dive Services (the company from which we were leasing Deep Rover) assigned to our expedition, had some of these traits, but not all. Mostly he had trouble multitasking and was deemed weak on "attention to detail" and "situational awareness." I was blithely unaware of his deficits, but others were not, which is why Etch had been brought in as part of the team. The original intent was for him to be with us for just the first few days, but as he became increasingly aware of Peter's deficiencies, he kept extending his stay. He pushed it as much as he could, but he had other obligations and eventually he was called to shore. It was the day after he left that my little misadventure occurred.

With Deep Rover, we performed pre-dive checks just as we had with Wasp, to make sure everything was operational and properly secured. For this dive, José ran me through the process. He called out equipment items and control systems and ticked off the boxes on his clipboard each time I said "Check." With the pre-dive complete, I reemerged for the last order of business, a trip to the head.

There are no bathroom facilities* in a submersible, so this is de ri-
gueur, whether you think you need to go or not. While I was in the
head, Peter made an adjustment to the sub that nearly cost me my
life.

The way it was relayed to me later was that there was a discussion
in which Robi pointed out a potential problem with one of the emergency procedures. If the sub ever became entangled in something
like a ghost net† or cables on a wreck, it would be possible to jettison
the battery, frame, thrusters, and manipulators to lighten the load
enough to hopefully break free. This "hull rollout" was an entirely
theoretical maneuver. Nobody had ever tried it, because the risk factors were too great. With all that weight removed, the buoyancy on
the sphere was such that it would rocket to the surface with enough
momentum to blast through the interface and go airborne. There
was no restraining harness, and not so much as a seatbelt on the
pilot's seat, so it seemed pretty clear that this "safety" system hadn't
really been thought through. However, it wasn't the lack of a harness
that Robi was pointing out; it was the fact that the hand lever we
were supposed to pump to actuate the jettison would physically bang
into the seawater inlet valve.

At this point, I would like to go on record as saying that I think
having a seawater inlet valve on a submersible is a bad idea. There's
a very good reason why Navy submarines don't have portholes: It's in
recognition of the fact that anytime you violate the integrity of the
hull, you are potentially allowing the pressure monster a point of
ingress. Ironically, the seawater inlet valve on Deep Rover is one of
its safety features. The idea is that if the sub ever becomes stuck on

---

* Strictly speaking there are "piddle packs," but since these were clearly designed
by men for men, I have made it a lifelong goal to never have to resort to using
one.

† Ghost nets are fishing nets that have been lost at sea. They entangle and kill
untold numbers of sea creatures every year, including whales, dolphins, sea turtles, sharks, and fish.

the bottom, there is enough life support for three days, but because of the limited payload, there is no drinking water reservoir. There is, however, a small desalinator, so you can open the valve, collect some seawater in a bottle, and then run it through the desalinator to remove the salt anytime you need a drink.

Peter agreed with Robi that the valve handle would be in the way and responded by saying, "No problem." He quickly removed the handle and placed it in the toolbox behind the pilot's seat. It all happened so fast that no one noticed that, in order to access the Allen screw that held the valve handle in place, he had turned the handle counterclockwise, opening the valve. When I emerged from the head, Peter failed to point out the reengineering that had occurred in my absence, and I climbed into the sub unaware of the open valve and its missing valve handle.

In this early version of Deep Rover, there was no hatch to climb through. Rather, the acrylic sphere was split into two hemispheres that were hinged at the top like a clamshell. You simply climbed through the gap at the bottom, then a crew member cranked the two halves together, sealing you inside. As with Wasp, there was no way to extract yourself, even at the surface. We joked about the need to have paper and markers inside so that if we ever came to the surface away from the ship, we could write instructions to our would-be rescuers about how to free us from our shell. This feature never bothered me when I was awake, but it occasionally plagued my dreams. Never before or since have I dreamed about being buried alive in a coffin, but I had that nightmare twice during this expedition, once so vividly that I woke up clawing at the bottom of the bunk above me. The lizard brain abhors entrapment.

As I flooded my ballast tanks, after I confirmed I had a seal, my focus was overhead, watching the water close in over the dome and then following the dancing trail of my ballast bubbles as they ascended in a shimmering stream. Each iridescent orb expanded with the lessening pressure as it rose upward, striving to rejoin its native sphere above the rolling ceiling of the sea. The scrubber fans, which were cleansing the air in my own personal bubble, whirred in the background, blending with and largely masking a faint, unfamiliar whine.

It was subtle, but once I convinced myself the sound was real, it seemed louder in my right ear than my left. Thinking it might be something electrical, I sniffed the air, but detected no hint of heated metal or burning insulation. Nonetheless, there was no denying that the noise was there, and growing louder. I was reporting my depth in fifty-foot increments over the through-water comms and had just announced passing 350 feet when my physical contortions aimed at locating the source of the noise caused my stockinged feet to slide down into the seawater that had been streaming in through the open valve.

The water was ankle deep. Its source was immediately obvious, but where the hell was the handle to turn it off? I blew ballast, jammed on the vertical thrusters, and then waited for what seemed like an eternity to see if I was already too late. Had I passed the tipping point? This was a classic example of a positive feedback loop, where the deeper I sank, the greater the pressure, and therefore the more water came in, making the sub heavier, causing it to sink faster, increasing the pressure, and so on, until I either sank below my operational depth limit and imploded or hit bottom and drowned.

The submersible seemed to shudder briefly, and then, slowly, it began to rise, but with water still streaming in. I called the surface and had a little *WTF?!* exchange with Peter. I told him there was water coming through the seawater inlet valve and the valve handle was missing, and he told me it was in the toolbox behind my seat. I dug around frantically, pucker factor* increasing from 8 to 9 as I at-

---

*Military slang referring to adrenaline-induced tightening of the anal sphincter. The scale is 1 to 10, with 10 generally reserved for explosions, extreme disfigurement, or death.

tempted to locate the toolbox by feel. Once I did, I had to find the handle and its Allen wrench, then slide it over the valve stem and secure it with the set screw. When I had the handle in place, I tried turning it clockwise as hard as I could. It wouldn't budge. The pressure monster had his finger in the valve and wasn't giving up.

As the sub rose slowly through the water, the pressure decreased, allowing the air in the ballast tank to expand and gradually accelerating my rate of climb. Normally, I would have paused my ascent at fifty feet to make sure I wasn't going to come up directly under the ship, but I blew through that protocol, my only concern being to get to the surface as fast as possible. As soon as I broke through the surface, the Zodiac appeared with divers who immediately attached lines that hauled me to the side of the ship so they could attach the crane hook.

The team recovered the sub and had me back on deck in record time. When they cranked the sphere open, gallons of water came rushing out in a whoosh. However, a quick inspection revealed that the water level had stopped short of the electronics in the seat base, so, after things were dry and the seawater inlet valve had been changed out, the sub was ready to dive again in less than an hour. It was a get-back-on-the-horse moment. I was rattled, but still game.

This time as I flooded my ballast tanks, I was not reveling in my bubble stream or thinking of anything except basic survival. So, as I approached three hundred feet again, adrenaline still pumping, I was on high alert when I heard a very different sound from the whine associated with the leak. This was a series of long, piercing whistles, each increasing in pitch and volume. I looked around frantically inside the sub, then quickly realized the sound was coming from outside. It was an orca, nearly twice as long as the sub. It seemed to be checking me out. The distinctive white-on-black markings, the tall, bladelike dorsal fin, the streamlined shape, and the ease and power with which it cruised in from behind me, then circled the sub in slow motion before leaving, all heralded its position as apex predator in these waters. What a sight, and what a reminder of how poorly adapted we are to probe these depths compared with these distant air-breathing cousins.

Despite our limitations, we were making progress, learning new things about life in the midwater with every dive. Etch returned to oversee the remainder of our dives, which made us all breathe a lot easier. Peter never again worked as a submersible operations coordinator. The light wand was still producing no results, but, based on the complexity of the bioluminescence I was seeing and the apparent extreme need for stealth, it was becoming clear that this was too simpleminded an attack on the problem. On the other hand, the video recordings I was making of stimulated bioluminescence using the transect screen, which we now referred to as the SPLAT screen,* were far exceeding my expectations. I was starting to develop a library of bioluminescent signatures—recognizable spatial and temporal patterns of different bioluminescent displays that allowed me to identify the emitters. And many of these fantastic light shows, especially those produced by fragile gelatinous life-forms, had never been seen before.

We were also finally able to share the thrill with people who had never been in a submersible—not just scientists but the public. Near the end of the expedition, both CBS and NBC sent news crews out to the ship to report on our explorations. In its national news coverage, CBS included a sequence of my video recordings of bioluminescence. There was also a documentary made about our adventure, produced for the BBC, called *Dive to Midnight Waters*. It, too, showed my bioluminescence recordings, along with what at the time was some of the highest-resolution video ever recorded of deep-sea life. Being able to take the public along on our dives gave us all a huge

---

* Initially, *splat* was merely descriptive. It was many years later that my postdoc at the time, Sönke Johnsen, turned it into an acronym by coining a name for our process: the Spatial Plankton Analysis Technique.

sense of accomplishment, and we had high hopes that it might help stimulate additional funding for deep-sea exploration. Alas, it's just not that simple.

Seeing one news story about the deep sea or watching one or more documentaries is just a flash in the proverbial pan, with little lasting impact. Even when the public is transfixed by a subject, their interest does not translate to support. The space program arose not from public interest but from political interests. The perceived need to beat the Soviet Union into space resulted in NASA receiving a blank check in the 1960s, and it used some of that funding to create one of the best advertising and marketing campaigns ever seen for communicating science to the public. Public interest grew out of that campaign, which promoted space exploration as a fantastic frontier adventure tale complete with space-cowboy superheroes.

Funding for deep-sea exploration in general and bioluminescence in particular has always been a tiny fraction of that for space exploration. In fact, the only reason my interest in bioluminescence turned out to be fundable was that the Soviet Union was also interested in it. Had that not been the case, I doubt my adventures would have been possible.

Chapter 7

SEAS SOWED WITH FIRE

The ocean hides many secrets. But for those in the know, bioluminescence can reveal things otherwise unseen. It all boils down to
being able to read the light.

Ancient navigators could read the sea in the same way that the
Inuit people read snow.* Route finders drew on not just a lifetime of
study but *many* lifetimes, passed on from one generation to the next.
This knowledge was central to transforming the seas from hin-
drances into highways for exploration and opening up new frontiers
for settlement and trade. Knowledge was power, which is why, in
some ancient cultures, navigators were revered as priests and their
knowledge was zealously guarded as state secrets.

This secretiveness and a dependence on oral tradition mean that
most of the ancient knowledge is lost forever. A notable exception
was made possible by sailor and Polynesian scholar David Henry
Lewis, who tapped into the oral tradition of Pacific Islanders while it
was still extant, interviewing and sailing with South Seas navigators
to garner their secrets. His research, published in 1972 in a book
called *We, the Navigators: The Ancient Art of Landfinding in the Pacific*,
sheds light on a puzzle that had mystified many early European ex-

---

*A skill that allows them to track game in blowing snow, through slush, and over
hardpack.

plorers: How was it possible that these "primitives" in their outrigger canoes routinely crisscrossed a vast blue expanse that encompasses nearly a third of the Earth's surface area, managing to locate tiny specks of land without the benefit of such seemingly indispensable navigation tools as compasses, sextants, or maps?

Based on Lewis's findings, it appears that many factors were brought into play to achieve such remarkable maritime feats. Pacific Islanders steered by the stars, by the sun, and by the prevailing winds and swells. They followed the migration routes of tropical seabirds, tracking long-tailed cuckoos from Tahiti to New Zealand and golden plovers from Tahiti to Hawaii. They also expanded their target search, not depending on spotting a nearly invisible dot of land on the horizon but instead looking for clouds massing over islands and training dogs called *kuri* that barked when they smelled land.

Yet another extraordinary technique involved learning to read bioluminescence in the form of a phenomenon they called *te lapa*. Unlike ordinary bioluminescence, or *te poura*, which was seen at or near the surface, *te lapa* was deep luminescence, described as underwater lightning darting to and fro as streaks and flashes seen anywhere from a foot to six feet below the surface. The best navigators could gauge the distance to land from the patterns of the displays. Far from land, the motion of the "lightning" was slower than it was close to the land, where it was more rapid and jerky. It was most visible eighty to a hundred miles out and essentially disappeared eight to nine miles offshore. Navigators also claimed to be able to distinguish reef *lapa* from land *lapa* because the light from reefs was slower-moving than that from equally distant islands. Lewis personally observed both reef *lapa* and land *lapa* and claimed that they were readily distinguishable.

Te lapa, which Lewis first learned about from Polynesians in the Santa Cruz–Reef Islands, was described almost identically by Micronesians in the Gilbert Islands, who called it *te mata,* and by Polynesians in Tonga, who called it *ulo aetahi,* which translates as the "glory of the seas." The fact that this unusual method of land finding was shared by remote indigenous peoples suggests that it may have been a common part of the South Seas navigators' tradecraft. But what was it, exactly?

Lewis suggested that swells and backwash waves reflected off islands might play a part. Just like sound waves, ocean waves bend (refract) and reflect off solid objects, and in the presence of multiple islands they can create distinctive swell interference patterns. Although South Sea Islanders did not have charts in the traditional sense, they did create stick charts, constructed of palm ribs tied together with coconut fibers and with cowrie shells used to represent islands. These charts were not created to represent true distances, but rather were intended as mnemonic devices to remind the user which swell patterns were associated with which specific island groups.

The bending of waves around solid objects doesn't happen only at the surface. It can also occur in deep water because of the presence of what are known as internal waves. You can visualize such waves in a full bottle of salad dressing, where lower-density oil sits atop higher-density vinegar. Tilt the bottle back and forth and you can create an internal wave where the oil and water meet. In the ocean, such layers occur when warm water overlays colder, denser deep water. In places where reefs or other projections protrude into these layers, it's even possible to have internal breaking waves that create turbulence at the interface. Whether or not this helps explain *te lapa* remains to be shown.

Being able to see *te lapa* for myself and, better yet, film it, is on my bioluminescence bucket list. The closest I have come to anything comparable was once when David and I were on a camping/kayaking trip in the Sea of Cortez. We were night paddling under a full moon. The water was so clear and calm that it seemed to disappear beneath our boat, and I felt a sense of vertigo looking down at the moonlit bottom many feet below, where flashing dinoflagellates swirled and danced in an unseen current. It did not look like under-water lightning, but evidently it was the result of mechanical stimulation created by the turbulence of a bottom current interacting with the rock-littered seafloor.

It seems unlikely that *te lapa* involved prolonged interaction with the seafloor, since that would have been evident and therefore mentioned. Studies done by former fellow grad student Mike Latz and his colleagues have demonstrated that the conditions needed to stimulate bioluminescence in dinoflagellates are found under three common circumstances. The first is at the boundaries of moving objects, like boats or swimming animals; the second is when there is lots of turbulence, as in breaking waves; and the third results from flow at ocean boundaries, such as a current dragging water along the seafloor. The nature of the stimulus field responsible for *te lapa* doesn't fall easily under any of these headings and remains a mystery, but my money is on internal waves.

Although there have been quite a few experiments designed to pinpoint the nature and strength of the stimulus needed to excite bioluminescence in dinoflagellates, there is still controversy on this front. Rather than get embroiled in hydrodynamicists' nerd rants, however, I prefer to revert to my physiologist roots and think in terms of what I observed back in my graduate student days, when I was using a glass probe to jab single cells of *Pyrocystis fusiformis*. Under those circumstances, the most effective stimulus was one that caused a rapid deflection of the membrane. The question is, what kinds of disturbances are occurring with *te lapa* that might generate similar membrane distortions?

A lot can be learned about the kinds of stimuli that excite bioluminescence by observing a dolphin swimming through bioluminescent plankton. For a long time, the only way to record such observations was either verbally or artistically. The famed graphic artist M. C. Escher, best known for his lithographs and woodcuts of impossible constructions, also did more realistic works, including one entitled *Dolphins in Phosphorescent Sea.** This 1923 woodcut is a white-on-black depiction of the brilliant bow wave of a ship at night as it plows through bioluminescent plankton, with dolphins swimming in front of it. Each dolphin is outlined by the bioluminescence it excites, from the tip of its nose to the tip of its tail, which leaves an undulating luminescent trail. There are also two dolphins, one to either side, that are leaping clear of the water; one is creating a forward-directed corona of spray as it reenters the water, and the other is creating a backward-directed corona as its tail splashes down.

This depiction was at odds with many published scientific descriptions that reported an absence of bioluminescence on a dolphin's body. However, as it turns out, Escher got it right. With the recent advent of low-light-sensitive cameras, it has been possible to record scenes very similar to the one he depicted. Analysis of these videos reveals that the brightest light is associated with the spray generated when the dolphin breaks the surface. There are brilliant trails of light coming off the fins and tail (spots where flow transitions from laminar to turbulent†), but there is also light visible everywhere on the dolphin's body. The reason this was not reported by many science observers may be that they were not sufficiently dark adapted to see the dimmer bioluminescence associated with the sleek dolphin torso. In any case, the key finding from such analyses is that the amount of bioluminescence seen is simply a function of the volume of water being stimulated at any given moment.

Since the bioluminescence that is stimulated is a consequence of both the shape and the swim pattern of the swimmer, different fish

---

*According to the *Scientist's Guide for How to Win Friends and Influence People*, you should take every opportunity to point out that Escher's title is incorrect and it should be renamed *Delphinus delphis in Bioluminescent Sea*.

†An example of such a transition between different kinds of flow can be seen in smoke rising from a cigarette that starts out smooth and laminar, where the layers don't mix, and then becomes turbulent and chaotic . . . so it can insinuate itself into every fiber of your clothing.

have different bioluminescent signatures. This is a characteristic that nighttime fishermen exploit. I learned of this from one old-timer who used to fish the Florida waters near where I live. He described how back in the day, when dinoflagellate bioluminescence was so abundant in the local estuary that the fishermen described it as "fire in the water," they could identify the fish by the luminous patterns they made.

If fish are easily spotted and identified by the distinctive patterns of light they produce, then it should be no surprise that the same is true for ships and submarines. In fact, because of the enormous volumes of water they excite, some of these trails can be seen from great heights above the ocean.

Astronaut Jim Lovell, best known as the commander of the ill-fated Apollo 13 mission, could bear witness to the visibility of bioluminescence from on high, based on a close call he once had as a U.S. Navy pilot flying Banshee night fighters in 1954. He had been on a training mission off the aircraft carrier *Shangri-La* in the Sea of Japan. He finished his flight and had begun following what he thought was the homing signal leading him back to the carrier when he realized he was heading in the wrong direction. The signal he was tracking was coming from mainland Japan and coincidentally broadcasting at the same frequency as the carrier's. Once he realized his mistake, he needed to try to communicate with the carrier. To do so required reading the communication codes written on the pad of paper strapped to his thigh.* The trouble was, the cockpit lights were too dim to read by, so he flipped on a small light that he had jury-rigged into the airplane's electrical receptacle. The result was a short circuit—a bright flash of light followed by a total blackout of all of his instrument readouts. Disaster!

Without his instruments, he had no hope of locating the carrier. As he sat in the darkness, contemplating the grim truth that if he

* Pilots call this a kneeboard.

had to ditch, his odds of survival were nil, he desperately scanned the sea below for any sign of the carrier. It was like trying to find a needle in a haystack . . . *in the dark*. But in the end, it was the darkness that saved him, because the total blackout allowed him to spot a faint shimmering trail of light in the water. It was the bioluminescence stimulated by the long, turbulent wake the carrier left behind. He recognized it for what it was—his illuminated path to salvation—and followed it all the way back to the *Shangri-La*.

Being able to see bioluminescence churned up on the surface by an aircraft carrier from an airplane is one thing, but how about that which is stirred up by a submarine from a satellite? Whether or not this was actually done I can't say, but it was theoretically possible, and definitely doable from P-3 Orion airborne submarine hunters, which is why the U.S. Navy cared and why my fascination with bioluminescence turned out to be fundable.

During the cold war, submarines became critical for intelligence-gathering efforts on both sides. Most stories of the behind-the-scenes drama associated with playing cat-and-mouse games underwater remain classified, but some of these were revealed to the public for the first time in the book *Blind Man's Bluff*, by Sherry Sontag, Christopher Drew, and Annette Lawrence Drew.

The audaciousness of the missions was breathtaking. They included sneaking along Soviet coastlines, periscope up, looking for posted signs that were the Russian equivalent of DO NOT ANCHOR—CABLE HERE and then following the cable offshore and securing an eavesdropping device to it. They also involved learning to acoustically track increasingly quiet Soviet missile subs called boomers that could sneak up on our coast and unleash nuclear Armageddon in the form of ballistic missiles.

The prime directive, for both the U.S. and Soviet submarine fleets, was "Avoid detection at all costs." Staying submerged and quiet was key to accomplishing this, but it wasn't a guarantee, which is why both sides were pursuing less traditional modes of detection that fell under the heading of nonacoustic anti-submarine warfare (ASW). This was not a new idea. In fact, during World War II, when German U-boats were slipping into the Gulf of Mexico and torpedoing freighters within a hundred miles of Florida's coast, the Germans were well aware of the threat of detection that bioluminescence posed. One U-boat commanding officer, Captain Reinhard Hardegen, deemed it the primary threat and warned fellow commanders, "The most dangerous feature of American water is marine phosphorescence at night—because of aircraft or destroyers, be aware that if you travel at periscope depth, vortices off your screws and cannon will show up as phosphorescence and betray your position."

Because bioluminescence could reveal the presence of submarines, both the Soviets and the U.S. Navy put significant effort into developing a predictive capability for the phenomenon, in order to know where and when they or their enemy might be most vulnerable to detection. Initially, this didn't seem like any great challenge.

Measuring bioluminescence is easy, as the first investigators who dropped light detectors into the ocean discovered. Just lower a sensitive enough light sensor over the side of a ship and wiggle it around and you'll record bioluminescence. In fact, shortly after investigators recognized that the amount of light they were measuring was related to the sea state, they started designing systems that permitted greater control of the stimulus. All of these were called bathyphotometers (BPs), which simply means "deep" (*bathy*) "light meter" (*photometer*).

BPs came in a great many shapes and sizes, but the most typical design used a pump to pull water into a darkened chamber, where a spinning propeller or a narrow constriction stimulated bioluminescent critters to flash so their light output could be recorded. The problem was that the light measured depended on the size of the detection chamber, the method of stimulation, the flow rate, and how much the water was churned around before it got into the detection chamber. This meant that numbers measured by one kind of

BP couldn't be compared with those measured by another. Also, most BPs had low pumping rates (a liter per second or less), which raised concerns that the only kind of bioluminescence they were measuring was probably from dinoflagellates and not other, faster and much brighter emitters, such as krill, that could easily evade such wimpy flow fields but might contribute significant light when colliding with an onrushing submarine.

All of these various concerns came to a head in 1981 when, at the request of the oceanographer of the Navy, a panel of university and Navy experts convened to propose improvements to BP design. Based on their recommendations, a request for proposal (RFP) was issued to see who might come up with a design that could address all the Navy's concerns and become the U.S. Navy standard system for measuring bioluminescence around the world.

Jim Case submitted a proposal, and, since much of what he was advocating was based on my thesis research, he included me as co–principal investigator. Being a co-PI on a grant of this magnitude so early in my career was a huge deal. This was big money—over half a million dollars to start with, and more to follow—and potentially high-profile, but only if it worked. According to the rules of credit and blame in research, success is attributed to the supervisory brilliance of the adviser, and failure to the utter incompetence of the grad student. That formula goes double for postdocs. If this project failed, it could be reputational suicide. Therefore, when I learned we'd been awarded the grant, I felt a weird mixture of glee and angst—sort of like the guy who finds out he won the lottery and then realizes he's misplaced the ticket.

Although Case and I were nominally co-PIs, he had his hands full serving as UCSB's associate vice chancellor of research, which meant the brunt of the project fell to me. I had experience with instrumentation development and a solid enough background in the science needed, but I had never managed a project of this scale. There were many personalities and moving parts, and at numerous points along the way I thought the whole damn thing was going to wind up in an epic fail.

Our guiding principle was to have the biology drive the engineering. To make sure commonplace fast swimmers couldn't escape measurement, we used the top swimming speeds of krill to calculate what the pumping speed needed to be. The trouble with that requirement was that the faster the flow rate used, the shorter the residence time of the light emitter in the chamber, which meant that with many animals, we couldn't measure their entire flash. If we used a standard-size detection chamber, a krill would go rocketing out the exhaust before we'd had time to measure more than a fraction of its light output.

To get around this problem, we made the detection chamber into a tube over four feet long and almost five inches in diameter. That created three more challenges: how to force water through the tube at high speed, how to stimulate the bioluminescence in some calibrated fashion, and how to measure it (also in some calibrated fashion). We settled on a high-speed pump at the back end of the tube that pulled water in through a steel grid at the front end, thereby producing a well-defined plane of stimulation—sort of like a SPLAT screen but coarser.

In order to collect all the bioluminescence in an unbiased fashion down the length of the tube, we embedded more than seventy fiber optics that collected the light and directed it to a photomultiplier tube. We also designed a freely rotating light baffle in front of the grid to keep out stray light from the moon or ships, while minimizing pre-stimulation of bioluminescence before the water hit the stimulation grid. This feature was deemed especially important because of what I had learned during my thesis research on *Pyrocystis fusiformis* about how much brighter the first flash is than subsequent flashes, which meant that our measurements would be drastically reduced if the sources were stimulated to flash before entering the test chamber.

Since this was a Navy project, it required a Navy-worthy acronym. I struggled for a while before settling on one I was happy with: the High Intake Defined Excitation Bathyphotometer, or HIDEX-BP, was a name that incorporated all the system's most original features. I thought it had a nice ring to it. The lead software engineer on the project, Steve Bernstein, a.k.a. Bernie, disagreed. He thought it sounded too much like Riddex pest repellent, which is why, shortly after I came up with the name, the control screen for the HIDEX software started displaying pop-ups advertising "HIDEX for all your pest control needs." Theoretically, these were removed before the system was delivered to the Navy, but, given Bernie's sense of humor, I was never completely confident they wouldn't reemerge at some later date.

The key to designing a good measurement system is being sure you know what the numbers mean. Meeting this challenge was daunting and involved many steps, one of which found me dangling over the edge of a large fiberglass septic tank (new, not used) with the top cut off in order to create a test tank big enough to hold our prototype BP. The edge of the fiberglass was cutting into my chest, and I was sweltering under the big sheet of black plastic that covered the tank, to keep out stray light, while my arms were going numb in the chilly seawater, where I was slowly releasing bioluminescent di-noflagellates into the mouth of the BP. This was one of many points along the way to developing the HIDEX where I found ample opportunity to reevaluate my career choices. In no way did this re-semble the popular view of the everyday life of a marine biologist— swimming with dolphins during the day and sipping mai tais on a tropical beach at sunset.*

The initial field test for the HIDEX was my first experience serv-

---

*For a more accurate view of the true nature of life as a marine biologist, see Milton Love's classic rant "So You Want to Be a Marine Biologist?" (*Science Creative Quarterly*, September 28, 2007).

ing as chief scientist on an expedition, and although I had plenty of familiarity with Murphy's law from previous seagoing adventures, I felt like Murphy outdid himself on this one. The seas were high, the vessel unstable, and on the first night, the combined smells of eau de diesel and rancid oil from fish frying in the mess left most of my team seasick. Once we finally got enough operational hands to deploy the HIDEX, it didn't work. When we troubleshot it back on deck, the consensus was that it didn't run when wet or in the dark. Rather than report this state of affairs back to Dr. Case, we agreed that the better course of action would be to force the ship's captain to take us to Baja, where we would open a bar and hang the HIDEX out front, disguised as a beer keg, with a sign reading HOT BEER, LOUSY FOOD, BAD SERVICE, HAVE A NICE DAY.

Battlefield humor, usually at one A.M., when everyone was punchy and nothing was working, became one of the hallmarks of the brutal campaign to get the HIDEX operational. It was very much a team effort that required maintaining a "plays well with others" factor, which we managed pretty well most of the time. Nevertheless, this was a big project that went through the standard phases of (1) enthusiasm, (2) disillusionment, (3) panic, hysteria, and overtime, (4) hunt for the guilty, (5) punishment of the innocent, and (6) reward for the uninvolved.* The by-product for me was unrelenting stress, only partially eased by a steady diet of Tums, which eventually escalated to chugging Maalox straight out of the bottle.

Ultimately, though, HIDEX was deemed ready for its first Navy mission. This was to be a transatlantic crossing, making bioluminescence measurements in the top five hundred feet of the ocean from the Canary Islands, off the northwest coast of Africa, to Florida. There were five of us making the crossing: me, Dr. Case (whom I was now allowed to call Jim), Bernie, Mike Latz, and our electrical

---

*Wikipedia, s.v. "Six Phases of a Big Project," en.wikipedia.org/wiki/Six_phases _of_a_big_project.

engineer, Frank. It was a classified mission to be carried out aboard a Navy oceanographic vessel, the 285-foot USNS *Kane*.

At the outset of the HIDEX project, I had been issued a security clearance and was familiar with the restrictions this entailed, but this mission brought a new level of security that struck me as over-the-top. While I was allowed to tell my husband the name of the ship and the name of the port we would be leaving from, for some reason I was not permitted to tell him both things on the same day. I made the security officer repeat that instruction just to be sure I heard it right. Apparently it was a carryover from World War II, when concerns about intercepted transmissions drove the instigation of precautions about not including multiple pieces of key intel in the same missive. It was difficult for me to imagine how any of what we were doing could possibly be of interest to a foreign power, but it was.

This interest became evident soon after we boarded the *Kane* in the port of Las Palmas. I was in the ship's lab, talking to Bernie, when the captain came in and, gesturing toward the HIDEX on the back deck, said, "You should probably throw a tarp over that thing. It's attracting too much attention." We walked out onto the fantail to see what he was talking about and were astonished to discover a huge Soviet oceanographic vessel tied up directly behind us at the dock. Up on its bow, a couple of guys with cameras and telephoto lenses were taking pictures of our baby.

Our next hint of more than casual interest came with the temporary disappearance of the shipyard's welder. Since the *Kane* didn't have a qualified welder on board, we had to use the shipyard's guy to weld the BP's winch onto the deck. At one point we lost track of where he was, and I was surprised and slightly alarmed when I discovered him in the lab, leaning over Bernie's shoulder and pointing to the HIDEX software, asking questions.

Although we might have been able to persuade ourselves that both of these incidents had innocent explanations, the next turn of events made such an interpretation more challenging. On a Navy vessel, the radio shack is classified space, and only those with appro-priate clearance are allowed entry. The night before we sailed, our radioman went ashore for some R&R and ended up going on a bar crawl with a few of the Soviet ship's crew members. Copious alcohol consumption led to the Soviets convincing him to invite them aboard the *Kane* and into the radio shack. In the debriefing that followed this incident, the seaman claimed that they had subsequently invited him aboard the Soviet vessel, but the last thing he said he remem-bered was walking up their gangway. When he came to, the police were fishing him out of the harbor. Our team was unaware of any of this until the following morning, when we saw our radioman being led away in shackles.

This was no minor occurrence, especially in the captain's view. He was convinced the whole thing was going to be a major blemish on his record, and he blamed us and the HIDEX. Therefore, while we waited for a new radioman to replace the one we had lost, the captain took the opportunity to hold an all-hands-on-deck meeting to talk about recent events.

Given my experience with U.S. Marines during David's tenure as a hospital corpsman, I had more than a passing familiarity with use of the F-bomb. For Marines, it's a form of punctuation often em-ployed where most people might insert a hyphen or an exclamation point. For the captain of the *Kane,* its use attained new, previously unimagined heights. In retrospect, I wish I had kept count, but when he started, it never occurred to me that we might be entering *Guinness World Records* territory.

The expedition was not fun. Besides rough seas and bad food, the captain was a jerk* who made it abundantly clear he was not effing pleased by my steadfast refusal to stop by his cabin to see his collec-

---

*He was not a U.S. Navy captain, but part of the civilian service that the Navy employs to man its oceanographic survey vessels.

tion of memorabilia—all supposedly engraved with his life's philosophy: SHIT HAPPENS. Nevertheless, the HIDEX performed well. It satisfied all of the Navy's requirements. We were issued a patent and it was officially deemed the U.S. Navy standard for measuring bioluminescence in the world's oceans. More important, I was able to kick my Tums and Maalox addictions.

As for the Soviets' part in the story, they figure into two interesting addenda. The first was a scientific paper that came out a year later in the journal *Oceanology*, in which much Soviet science research was published. The publication by several Soviet scientists touted a whole new kind of bathyphotometer, one that on the outside was a dead ringer for the HIDEX but on the inside bore no resemblance—apparently despite their best efforts, the only real intel they gathered was those photos they got the first night.

The second occurred at an International Symposium for Bioluminescence and Chemiluminescence that happened after the fall of the Soviet Union. I was sitting at lunch with a British colleague when several former Soviet scientists sat down opposite us. My colleague knew them and introduced me as the inventor of the HIDEX-BP. The gentleman opposite me did a classic jaw drop and then blurted out, "But I thought you were . . ." before he stopped himself. I'm not sure what he was going to say. A man? Older? Dead? There's no way to know for sure, but based on the expression on his face, it seemed to matter to him, which I decided to take as a compliment for the HIDEX, if not necessarily for me.

For the Navy, the HIDEX-BP provided the numbers they needed for calculating something called nighttime water-leaving radiance. If a submarine was moving through a particular ocean at night, at a certain speed and depth, how much bioluminescence would it stimulate, and was it detectable at the surface?

For me, the HIDEX opened up a new perspective on the nature of the bioluminescent minefield. The most surprising discovery it produced was the existence of thin layers of intense bioluminescence, less than twenty inches thick. Their presence had obvious strategic importance to the Navy. But from the perspective of understanding marine ecology, it was part of what at that time was becoming a revolution in thinking about the distribution of life in the ocean.

Because net sampling provides a jumbled mass of life, for a long time there was a widely held impression that the ocean is like a soup, with everything mixed together. But there were a number of new sampling technologies, some acoustic, some optical, some mechanical, that were allowing scientists to look at life in the ocean on a finer and finer scale.

The closer we looked, the more we found what was called patchiness. Animals weren't evenly distributed, but were found in clumps. For example, when I got to dive a submersible with a SPLAT screen through those bioluminescent thin layers, I discovered that they were composed of a dense aggregation of bioluminescent copepods. These copepods (*Metridia lucens*) were apparently feeding on a layer of marine snow that had accumulated between two water masses of different density.

Understanding the nature and causes of patchiness is one of the grand challenges in oceanographic ecology. It takes a lot of energy to hunt, which means food must be sufficiently plentiful that more calories are consumed than lost. The fact that calculations indicate many predators can't be sustained by the average concentration of their prey in the ocean suggests that those predators must have some means of finding and exploiting dense patches of prey. Could bioluminescence be a means to that end?

There is some evidence that southern elephant seals exploit bioluminescence to locate nourishment. As the name implies, this is the heftiest of all seals and the largest marine mammal that isn't a whale, so the species needs a great deal of food. To fill their stomachs, elephant seals spend about ten months a year at sea, diving continuously, day and night, sometimes going nearly five thousand feet deep (about four Empire State Buildings) to forage.

Seals don't echolocate, but they do have beautiful big brown eyes that are more sensitive than human eyes, and among seals, elephant seals have the most sensitive eyes of all. Since their primary prey are lanternfish and squid, there are two schools of thought. The first is that the seals are homing in on light emissions of their prey, and the second is that they are seeing their prey by the bioluminescence that the fish and squid stimulate as they swim through the biolumines- cent minefield of planktonic luminescence. Since foraging forays of elephant seals target the boundaries between water masses, which also tend to be zones of high bioluminescence potential, the mine- field scenario seems more likely to me, especially since it doesn't depend on incessant maladaptive spontaneous emissions by the prey.

If you want to know what it's like to swim through a minefield of living light, you can find out by visiting a bioluminescent bay. There are several in Puerto Rico. The boat that carries tourists out onto the bay startles fish that dart away in all directions, creating shimmering thin contrails as the vessel itself generates a billowing wake of neon- blue froth.

When the vessel stops, dangle your legs over the side and you will be immediately rewarded with shimmering sequined boots—a scin- tillating aura that surrounds your limbs, enticing you to kick more and more vigorously until you are gleefully splashing like a toddler in the bathtub, creating a watery eruption of brilliant sapphire.

If you are lucky enough to be some place where swimming is al- lowed, then you can go ahead and dive in.* As you swim, you will be enveloped in a twinkling halo of glittering stardust. Wiggle your fin-

---

* As long as you aren't wearing sunscreen.

gers in front of your face and watch sparks fly off your fingertips. It'll feel like you have been magically granted superpowers. You have! Life surrounds us and nature is everywhere, but too often invisible to our eyes. Here, the hidden energy of life is revealed and the response is universal—a heady fusion of joy and awe.

PART II

TO KNOW THE DARK

To go in the dark with a light is to know the light.
To know the dark, go dark. Go without sight,
and find that the dark, too, blooms and sings,
and is traveled by dark feet and dark wings.

— WENDELL BERRY, "To Know the Dark"

Chapter 8

GLORIOUS PUZZLES

It was a gulper eel. I'd never seen one alive before, but there was no mistaking that long, skinny, scaleless body and the enormous toothless mouth with jawbones nearly a quarter the length of the fish. It was swimming fast, with a snakelike undulation that kept it just ahead of the submersible. But it couldn't sustain the sprint, and after less than a minute it fell back and then suddenly stopped swimming. The sub pilot, Phil Santos, began maneuvering the sub to keep the fish in front of us as I adjusted the pan and tilt on the external camera, trying to get the gulper into its field of view. I looked down at the controls for a second, and when I looked up I couldn't grasp what I was seeing. Instead of the black fish, I saw a brown balloon on a black string. And then the balloon split open, as if along a seam, and simultaneously shape-shifted and color-morphed back into the form of the black fish before it started swimming again.

"Did you see that?" I yelled to Phil, wondering if my eyes were playing tricks on me. Obviously, he had to have seen it, since he was scope-locked on the fish and now giving chase. We were at the end of our dive. The surface had already called us up, so there wasn't time to screw around with lining up the perfect shot. I gave up on the pan and tilt controls and grabbed the camcorder out of my satchel. The next time the fish stopped and performed its fantastic metamorphosis, I was filming. It opened its jaws ridiculously wide, and then, as it closed them, the thin brown skin of its gullet, which appeared black when deflated, expanded, creating a brown balloon that inflated even farther as Phil and I stared in utter amazement.

The scientific name for this fish, *Eurypharynx pelecanoides*, refers to its crazy mouth: the long (*eury*) pharynx (the membrane-lined cavity behind the nose and mouth) with pelican-like proportions. It is assumed that this extreme adaptation evolved to engulf prey, like a pelican, but nobody knows for sure because no one has ever seen it feed. It's rare to find these fish in trawl nets and much rarer still to see one from a submersible.

Undoubtedly, the hugely expandable mouth is used for consuming prey, but here was another possible function. "Is that to keep itself from being eaten?" Phil asked. I shrugged. "I guess. I didn't know they could do that. I don't know if anybody knew they could do that." I was speaking into the recording microphone, describing what I saw: "He puffed his jaw out and made himself into this big balloon." Then the eel did it again, this time on camera, and I hooted with glee. "That's incredible! There, we got it. Whatever the hell it's doing—I got it on video! Wow! That was so cool!" Then, in the midst of my exultation, just as it was deflating again, Phil gave the thrusters a kick, jockeying the craft in such a way that the fish slipped into one of the large collection cylinders on the front of the sub, and, before it could swim away, he activated the hydraulic lids, sealing it inside. I shouted, "Oh, man. You got him? Way to go, Phil!" It was an incredible feat—the equivalent of executing a perfect turnaround jump shot in basketball. I couldn't believe it.

On the audio track, I was still exclaiming, "That is amazing!" as Phil's deep, Boston-accented voice calmly announced, "Okay, topside, we're at 2,400 feet. Request permission to leave this depth." I tried to settle down to business, but my voice was shaking as I spoke into the recorder. "That is a gulper eel—I can't believe I'm saying it—in DS6.* Depth is 2,420 feet and temperature 4.2 degrees Celsius."† Not only had we been able to record its crazy antics on video, but Phil had captured it in a way that would give me an opportunity to study its bioluminescence.

More than anything, I wanted to get that fish out of the sampler and safely into a tank in the dark, because this was the best chance anyone had ever had to study its light-generating abilities. The gulper eel is a prime example of our general ignorance about how animals in the ocean use bioluminescence. It has an elaborate light organ on the end of its absurdly long tail, causing speculation that it might practice yogalike contortions to dangle its trailer light in front of its mouth as a lure. It also has a groove that runs down the length of its body like a racing stripe, which some scientists had speculated might also be bioluminescent, while others doubted it, and no one had a clue what function it might serve. But I was about to get a clue.

AFTER WE DELIVERED the HIDEX-BP to the Navy in 1989, it was time to move on from my extended postdoc position at UCSB to a real job. I knew I wanted to learn more about how animals use bioluminescence, which meant I needed to observe the light emitters. The options in this arena were limited. Although the use of remotely operated vehicles (ROVs) was on the rise, providing greater entrée to the deep sea, they were no good for the kinds of observations I wanted to make. They were noisy and obtrusive and essentially blind for my purposes, since the only way to see with an ROV is through

---

*The eight Plexiglas collection cylinders mounted on the front of the sub's lower work platform were known as detritus samplers (DS) because they were originally designed to collect detritus or marine snow.

† Using feet instead of meters for depth while using metric units for temperature may seem odd, but that's because those were the units used by the depth and temperature systems installed on the sub. For imperialists living in England, 4.2°C converts to nippy. For imperialists living in Florida, it's considered below the legal limit. For imperialists who prefer real numbers, 4.2°C = 39.6°F.

its cameras, and no cameras at that time were as good as the fully dark-adapted human eye. I needed a submersible, and not just any submersible; I needed one designed for the midwater, like Deep Rover.

At that time, there were fewer than a dozen deep submersible vehicles in active service nationwide. Of these, only five were being used routinely for serious scientific research, and of those, only two were designed for the midwater. Both were owned by Harbor Branch Oceanographic Institute (HBOI), in Fort Pierce, Florida. As luck would have it, in 1989 they were looking to hire an entry-level scientist with submersible experience. I applied and got the job, which allowed me to set up my own laboratory, designated, according to HBOI protocol, as the Bioluminescence Department. That made it sound a bit grander than it was, but it was unquestionably a plum job, most especially because it included access to HBOI's research ships and its submersibles, the Johnson-Sea-Links (JSLs) I and II.

The subs were named for the partnership of their financier and inventor, respectively. Because of his deep love of the ocean and its mysteries, Seward Johnson, Sr. (half of Johnson & Johnson), had committed part of his personal fortune to the ongoing development of these cutting-edge submersibles after their inventor, Edwin Link, had financed the early phase of their creation out of his own much smaller personal fortune.

Link originally designed the Johnson-Sea-Link as a means of carrying scuba divers into the ocean. The acrylic sphere where the pilot and scientific observer sat was not actually the business end of the sub; it was merely the bus driver's seat. The passengers in this "bus" sat in a separate compartment behind the sphere, an egg-shaped metal pod known as the dive chamber. The idea was to take the sub down to the limits of scuba diving depth, as much as a thousand feet, with both the sphere and the dive chamber maintained at atmospheric pressure. Once it was on the bottom, the pressure in the dive chamber was increased until it matched the outside pressure, allowing the downward hatch to fall open and the two scuba divers inside to swim out and get to work. It was the equivalent of performing an untethered spacewalk outside a spacecraft, but instead of the one atmosphere of difference between the inside of a spacesuit and the vacuum of space, scuba divers faced a pressure differential of as much as thirty atmospheres (440 pounds per square inch). Once they had exhausted their very limited bottom time, they could climb back into the dive chamber, close the hatch, and begin their decompression during the ascent. At the surface, the dive chamber coupled directly to a shipboard hyperbaric chamber where the divers would complete their decompression, which, for a dive to six hundred feet, with a bottom time of only four minutes, took *twenty-seven hours!*

Link soon realized that it would be more efficient if the collection of samples could be carried out with remotely operated devices, controlled from the front of the submersible. Over the years, he and his team of engineers developed an awesome array of collection systems: a robotic arm fitted with a variety of tools, including a claw, a benthic scoop, a suction hose, and cable cutters; a suction sampler called the "critter getter" that used a variable-speed pump and a carousel of twelve one-gallon Plexiglas buckets; and a much larger version of this same design with twelve three-gallon Plexiglas buckets. Samples collected with the claw could be dropped directly into these buckets, or they could be drawn into them through the suction hose on the robotic arm. And there were eight wastebasket-sized Plexiglas cylinders, called "D" samplers, like the one we used to capture the gulper eel.

ONCE BACK ON the deck, I made sure that the "D" sampler with the gulper eel in it was carried into the lab before anything else. The fish was still actively swimming, and as I slid the top open, I tried to figure out the best way to transfer it to a tank for observation. At well over a foot long, it was unwieldy, but highly flexible. I decided to scoop it up into a large glass finger bowl. The capture of this rare fish had drawn a small crowd in the wet lab. I lifted it out of the sampler and we all gasped in unison as a vivid neon-blue bolt of light flared along its length. Brilliant even under fluorescent lights, this was unquestionably the brightest, most dazzling bioluminescence I had ever seen.

The eyes of deep-sea predators are exquisitely sensitive—tuned to detect the very dimmest of flashes and often lacking defenses such as eyelids to block out bright light. To those eyes, a flash of that intensity would be devastating, like looking directly at the white-hot center of an electric arc welder with your eyes unshielded. Shape-shifting wasn't the gulper's only defense. Blinding an attacker was another option.

At least that was my best guess. An awful lot of what we think we know about bioluminescence in the ocean is just guesswork. A light organ next to the eye is a flashlight. A light dangled in front of the mouth is a fishing lure. But a light organ at the tip of such a ridiculously long tail and a brilliant racing stripe that runs the full length of the fish still leaves me suffering from a failure of imagination. And yet, what a glorious puzzle to try to solve: *What in the day-to-day life of a gulper eel makes these adaptations critical to its survival?*

Grasping what drives adaptation is fundamental to understanding the course of evolution. In today's world, figuring out how life can adapt in the face of rapid climate change is key to distinguishing potential evolutionary winners and losers and to knowing where to focus management efforts in order to minimize biodiversity loss. It's also critical that we recognize our own vulnerability. Believing that the world was designed to have us in it and therefore everything is going to be all right is a dangerous folly. I think Douglas Adams put it best:

> This is rather as if you imagine a puddle waking up one morning and thinking, "This is an interesting world I find myself in—an interesting *hole* I find myself in—fits me rather neatly,

doesn't it? In fact it fits me staggeringly well, must have been made to have me in it!" This is such a powerful idea that as the sun rises in the sky and the air heats up and as, gradually, the puddle gets smaller and smaller, frantically hanging on to the notion that everything's going to be all right, because this world was *meant* to have him in it, was built to have him in it; so the moment he disappears catches him rather by surprise.*

How is it that we or any other living creature came to be so well adapted to our environment? You can thank that old evolutionary two-step: *natural selection* acting on *heritable variation*. The transfer of information from one generation to the next is the basis of *being*. It is written in our DNA. Intriguingly, it is not a perfect transcription, but an imperfect one, because it is the imperfections—the experiments in *how to be*—that provide the framework for natural selection.

Nonlethal mutations produce offspring that can vary in form and function. You are not a perfect clone of your progenitors. There are differences. Some of these variations may prove beneficial to such a degree that your odds of surviving long enough to pass on your genes to your offspring are enhanced. The peppered moth is the classic example. During the industrial revolution, soot and pollution darkened the trunks and branches of trees favored by these moths, making the light-colored peppered moths stand out as easy prey for visual predators such as birds. As a result of increased predation on the light-colored moth, a darker-colored variant was favored, and in the course of less than fifty years it went from constituting just 2 percent of the population to predominating at 98 percent. Now that's success!

Marine bioluminescence is a comparable evolutionary success

---

* *The Salmon of Doubt: Hitchhiking the Galaxy One Last Time*, a posthumous collection of writings by Douglas Adams (New York: Ballantine, 2002).

story. Why? How did there come to be so many light makers in the ocean? The numbers are staggering. When William Beebe conducted trawls off Bermuda, in the same waters where he dove the bathysphere, he found that more than 90 percent of the fish he collected in his nets were bioluminescent. When you do the math, it turns out that we're not talking about mere billions or even trillions but possibly quadrillions of bioluminescent fish in the ocean.

If you measure success in terms of numbers, then bioluminescent fish are the most successful vertebrates on the planet. There are also shrimp and squid, as well as plankton (like dinoflagellates and copepods) and untold numbers of fragile jelly animals, that are part of this light-spangled bouillabaisse. Their numbers vary depending on depth and location, but in the open ocean, the largest living space on the planet, there is no question that bioluminescence rules.

Why has bioluminescence arisen so many times? Logically, it seems obvious that selection for light emission must have evolved after eyeballs did. One line of thought goes that the development of vision allowed the detection of prey at a distance, which led to an explosion of diversification—the result of an arms race between predators and prey. As the ocean filled up with predators that were becoming ever swifter and nastier, prey had to be able either to outswim their pursuers or to hide from them. In open water, where there are no hiding places, the only refuge was darkness. As a result, prey, followed closely by predators, began to move into darker waters.

Surviving at the edge of darkness favored certain mutations such as improved visual sensitivity. It also favored any camouflage that made prey harder to detect, such as countershading and counterillumination. Those that didn't counterilluminate would be more easily picked off by visual predators, just like the light-colored variants of the peppered moth.

Viewed in this light, it is not surprising that the small, unassuming bioluminescent bristlemouth fish is the most abundant vertebrate on Earth. Think of that: The numerically dominant animal with a backbone, numbering by some estimates in the quadrillions, is a three-inch fish whose most outstanding feature is the array of light organs adorning its belly—light organs that allow it to hide in a place with no hiding places. The conundrum is why the ocean isn't filled with nothing *but* this one kind of counterilluminating fish. Different species arise in places where populations get broken up and must adapt to separate circumstances. In a complex environment like a rainforest, a diversity of species makes sense: Animals occupying varied habitats exhibit distinctive patterns of coloration contingent on what background they blend with and unusual mouthparts depending on what food they eat. Once populations are separated, they change in singular ways, resulting in a diversity of forms.

Genetic isolation is the necessary precursor to the development of different species and the hallmark of natural selection. But in the open ocean, where there are no obvious barriers, why would there be any diversification? The answer is . . . sex.

The key to success in sexual reproduction is attracting more and better mates. Living in darkness makes it harder to locate mates, so once bioluminescence arose as a means of camouflage, its redirection for mate attraction provided added value and a possible route to genetic isolation. For example, deep-sea lantern sharks, besides having a dense array of tiny belly lights for camouflage, also have light-emitting patches on their flanks that look rather like tribal stripe decals on racing cars. These are species-specific, which means that while one species sports a long, thin lightning stripe, another is emblazoned with a scythe, while others have equally distinctive markings simplifying the process of recognizing a prospective mate. These differences apparently came about not because of physical barriers but because of sexual preferences.

A great example supporting the idea that bioluminescence has played a significant role in the speciation of the ocean can be seen by comparing lantern sharks with viper sharks, who also have belly

lights but no flank markings, and therefore no apparent sexual selection. While there is only one known species of viper shark, there are
thirty-seven species of lantern sharks!

The diversity of bioluminescent life-forms in the ocean is all the
more astounding when you compare it with the situation on land,
where light producers are exceedingly uncommon. Besides fireflies
and glowworms, there are also a few relatively rare click beetles,
earthworms, millipedes, some mushrooms, and one particular kind
of land snail. In freshwater, it's even more scarce—the only known
example is a limpet (a type of mollusk) found in the streams of
northern New Zealand. Clearly, these are exceptions rather than the
rule. The reason that luminescence is so uncommon outside the
oceans is thought to be that the existence of more and better hiding
places makes it so much easier for prey to hide from predators that
they don't need to depend on darkness.

Bioluminescence was well established in the ocean when life first
invaded land, lakes, and streams. But these early colonists were not
luminescent, so the ability to produce light needed to be reinvented—
something that was eminently doable, given how many separate
times it arose in the ocean. However, the selection pressure that
forced animals into darkness to hide was absent in a landscape
crammed with all manner of vegetation, and plenty of nooks and
crannies. If animals didn't need to live in the dark to avoid being
spotted by predators, then they didn't need to evolve bioluminescence to help them survive.

Retracing bioluminescence through evolutionary history is
greatly complicated by the fact that so little of it is preserved in the
fossil record. Photophores can sometimes be discerned in well-
preserved specimens of fish such as lanternfish and hatchetfish, but
in most cases there is no visible external manifestation in a fossil
that can be definitively identified as bioluminescent—so how can we
ever know for sure how bioluminescence arose? One way is to get
very, very lucky.

*   *   *

GETTING WILDLY LUCKY is exactly what happened on a mission in
the Gulf of Maine near the end of the summer of 1997. I was serving
as co–chief scientist along with my former postdoc and now collabo-
rator Tammy Frank, on my fourteenth expedition with the Johnson-
Sea-Link. We were diving in a place called Oceanographer Canyon,
along the southern rim of Georges Bank. Tammy and I had just returned from a daytime dive where we were running transects at different depths to document animal distribution patterns at different
light levels. Between transects, if we saw anything interesting we
would collect it, and on this dive we had spotted something quite
unusual as we were ascending from 2,600 feet toward 2,400 feet: a
strange-looking red octopus. When we first spotted it, the animal
was hanging inverted with its arms outstretched. Webbing between
its arms made it look like an open, upside-down umbrella. As we
drew closer, the creature first attempted to escape by executing a
jellyfish-like contraction, but after only one slow-motion medusoid
pulse, it gave up swimming and inflated itself into a highly distended
balloon. For several minutes it held that pose, and then it started
slowly flapping the two large fins on either side of its head in a scull-
ing motion while simultaneously twisting its body to deflate the balloon. That's when the pilot managed to capture it.

Back up on the ship, we transferred the football-sized octopus
into a large Plexiglas tank in the wet lab so we could observe and
photograph it. It was both bizarre and beautiful, and I was trying to
document it from every possible angle. Unlike every other octopus I
had ever observed in a tank, it did not affix itself to the sides or the
bottom, but instead hung in the center while contorting its elastic
body into a series of fantastic shapes. It had just spread its arms—
allowing me to get a great shot of its mouth and the underside of its
web—when my current postdoc, Sönke Johnsen, leaned over my
shoulder and said, "Those don't look much like suckers." I lowered
the camera so I could look with unobstructed eyes, and I had to

agree. They looked like shiny white pearls, or, more specifically, like photophores. This was a surprise, because while bioluminescence is common in squid, it's rare in octopods. In fact, there were only two known examples, and their light emission has nothing to do with their suckers: It emanates from a peculiar yellow ring with scalloped edges that surrounds the mouth of the female, and only at certain times—likely when they're trying to attract a mate. Bioluminescent suckers, however, were unheard of.

We immediately transferred what we were now calling the red balloon octopus* into a smaller container so we could carry it into a dark room. Sönke and I positioned ourselves on either side of the octopus, and then after Sönke turned out the lights, I prodded the creature gently with my finger. There was an immediate response as blue light blinked on and off from the sucker-photophores asynchronously, creating a lovely twinkling effect. This was a significant discovery all on its own, but later, when we were able to microscopically examine cross sections of the light organs, we realized it was even more momentous than we'd first imagined. We saw vestigial muscle rings characteristic of suckers. They were suckers that had evolved into photophores, which is why the photograph I was taking when Sönke leaned over my shoulder eventually ended up on the cover of the very prestigious scientific journal *Nature,* where we published our discovery. Here was an example of evolution caught in the act.

As with the peppered moths seeing a shift from light- to dark-colored variants in response to industrial pollution, many populations have undergone similar major evolutionary alterations due to changing environments. Charles Darwin said, "It is not the strongest or the most intelligent who will survive, but those who can best manage change." Put more succinctly, such changes force life to *adapt or die.*

---

*Its scientific name is *Stauroteuthis syrtensis,* but since our discovery, its common name has become the glowing sucker octopus.

Therefore, as visual predators proliferated, octopods needed to find a way to hide or they'd perish. Many populations adapted by becoming masters of camouflage, but a few, like the glowing sucker octopus, took a different tack and moved to deeper, darker waters. Dimmer light makes it harder to be seen by predators, but it also makes it more difficult to locate and attract a mate. Many octopods seduce mates by throwing their arms up over their heads and displaying their suckers as if they were in a wet T-shirt contest: "Hey! Look what I've got!" Under such circumstances, it makes sense that sexual selection would favor mutations that made the suckers more visible.

Because food is scarcer in deeper waters, once the suckers became more detectable, they also became valuable for a wholly different function—attracting prey—which helps explain how the glowing sucker octopus developed such an unusual diet. While most octopods consume things like scallops, crustaceans, and fish, the glowing sucker octopus survives solely on a diet of copepods. Since copepods are like the insects of the sea, this food regimen is the equivalent of a Florida raccoon living on a diet of mosquitoes. Sure, they are plentiful, but how can they possibly be gathered up in sufficient numbers to make a full meal? This is where the bioluminescent suckers come in. By hanging upside down in the water and twinkling its suckers, the octopus presumably looks like a yummy patch of plankton, which will attract copepods. Once a swarm has assembled, the octopus forms itself into a balloon,* sealing the copepods inside. Then it pulls its arms down toward its mouth, where a mucous layer traps the copepods, which can then be consumed like a gourmet seafood aspic.

With the development of glowing suckers that could be used for enticing both mates and food, this octopus variant could abandon its

---

* Behaviors can serve multiple functions, and it's likely that the octopus also uses its balloon inflation as an antipredator defense, as seen in the gulper eel's trick.

bottom-dwelling existence, where its suckers were useful for hanging on to things like rocks and shellfish, in favor of an open-ocean existence where their adhesive properties were unnecessary. Once a body part becomes obsolete, it is no longer favored in the selection process and gradually degenerates over many generations, as mutations that cause dysfunction will be selected against. This is how suckers become light organs, while still retaining some vestigial properties of suckers.

According to this hypothesis, bioluminescence in the glowing sucker octopus originally came about because of sexual selection. Tinder didn't invent sexual selection; it's been around since the invention of sex. But, as with Tinder, sexual selection has led to some pretty bizarre adaptations. The male peacock's tail is a particularly showy example. Could the gulper eel's tail or its racing stripe boast similar origins? How can we ever know, when no one has ever seen it use its taillight or racing stripe in its natural environment? If that taillight is a lure, used to draw in either food or a mate, then what kind of calculus is required to turn it on? How great is the risk of revealing itself to potential predators? And, if it does, what are its chances of escape using inflation or blinding? These are all questions that could be answered if only we could observe the fish in its natural environment, but how was that ever going to be possible? I wondered. It was a notion that kept recurring in my career, like an itch I couldn't scratch.

Chapter 9

STORIES IN THE DARK

T he drop-dead date was November 18, 1997. If the clearances didn't come through by then, the expedition was off. The day came and went, but I hardly noticed, because when I accepted the invitation to go, I didn't think it was really going to happen. What were the odds that Fidel Castro was going to allow an American oceanographic research vessel with a high-tech submersible into Cuban waters? This was, after all, the same Communist dictator whom the United States targeted for assassination no fewer than 638 times.* There were also attempts to destabilize his regime with an invasion, a counter-revolution, and an economic blockade that had effectively prevented American dollars and American tourists from entering the country since shortly after the 1959 revolution that brought him to power. The whole concept seemed preposterous. Nonetheless, if permission *was* granted, it would be an opportunity to explore previously inaccessible waters of the deep Caribbean. It would also be a free ride, because Discovery Channel was going to pick up the tab, so I said, "Sure."

The clearances were granted three days after the cutoff date, which clearly wasn't as much of a line in the sand as I had been led to believe, because suddenly the expedition was on again. I had done

---

*Castro claimed, "If surviving assassination attempts were an Olympic event, I would win the gold medal."

exactly zero prep and now there were only one and a half weeks until castoff. Tammy Frank, who at this point was an assistant scientist in my lab at Harbor Branch, was also going to be part of this expedition. It seemed like the two of us had just finished unpacking and stowing all the equipment used on the Gulf of Maine expedition, only two months prior, when we had captured the glowing sucker octopus. It was going to be an insane scramble to assemble and calibrate all the gear we planned to take.

As he has done so many times throughout my career, David stepped in to save the day, assisting with the last-minute packing, loading the gear onto the ship, and helping to get everything up and running in my shipboard lab. It drizzled all day on December 4, a very un-Florida-like thing to do, and as our early-evening departure time approached, I was already feeling deeply homesick.

Over the years, David and I have had plenty of practice being apart, but we hadn't been separated for Christmas since the first year we were married, when, six months after we said "I do," the Navy shipped him off to Guantánamo Bay, of all places. As a newlywed bride who was, frankly, astonished by how much I was loving married life, I was devastated to have my husband ripped away from me for six months. That early separation prevented me from ever taking him for granted. David evinced a similar response, first to almost losing me during my back surgery and then to having to accommodate my seagoing existence. *Never take each other for granted* is the kind of marriage counsel best rooted in experience.

Now as we stood on the dock saying our goodbyes and facing a separation of thirty-seven days, including Christmas and New Year's Eve, we fell back on a tried-and-true diversion: focusing on something fun in the future. We decided that since we had to be apart for our first and twenty-fifth Christmases as a married couple, with one and then the other of us in Cuba, we should plan to spend our fiftieth Christmas together in Cuba—with the hope that we're still alive and ambulatory and the travel ban has been lifted.

The political isolation of the destination wasn't the only thing that set this trip apart. There was also the fact that Discovery Channel was paying for everything. I had been involved with documentary productions in the past, but in those cases the filmographers had been guests of the scientists. This time it was the other way around, which was going to make for drastically different priorities.

Scientists are keenly aware that we need to find more and better ways to communicate the significance of our research to the public and that clearly television is a powerful means to that end, but most scientists distrust the medium. While hyperbole is used routinely for storytelling (think Babe the Blue Ox), it's generally contrary to science (think excruciatingly esoteric research paper).

One of our greatest and most remarkable characteristics as a species is our ability to pass on knowledge from one generation to the next. With the invention of the written word and then the printing press, radio, television, the internet, and social media, that ability has grown exponentially, making it possible to learn from vast numbers of people you will never meet. But there is a potential downside to this gift, which is that misinformation and lies can be passed on just as easily as truth. So how do we know what is true? Science provides the best solution to date.

The whole basis of the scientific revolution was the realization that it is possible to test for truth. The core concept, known as the scientific method, is that when you have a question about the truth of an idea, you need to form a hypothesis—a plausible explanation for your observations that is testable. To be useful, a hypothesis must be disprovable. Ideally, what you want to do is form multiple alternative hypotheses that could explain the observations you are trying to understand, and then systematically go about trying to disprove each one. The one that you fail to disprove is the most likely explanation—at least until more and better information comes along.

That's the key point: In science, you can never prove anything is permanently true. You need to always be open to alternative explanations should new information come to light. That means that, to be a good scientist, you *must* be comfortable with doubt, which makes speaking in absolutes difficult and giving yes-or-no answers to what seem like simple questions nearly impossible. But "It's complicated," followed by a long-winded explanation, heavily encumbered with qualifiers, is generally antithetical to good storytelling.

Just as the scientific method revolutionized how humans think about our world and our place in the universe, it can transform how we cope with the fire hose of misinformation being spewed at us in the so-called information age. But that requires that we do a much better job of teaching science as a way of *discerning truth,* which would be greatly expedited if scientists and television producers could find the right balance between science's quest for truth and television's demand for entertainment. That balance hinges on establishing trust between the players. Our Cuba expedition provided an object lesson in how *not* to do it.

THE PROBLEMS BEGAN before we even left the dock. A meeting on the bridge of the 204-foot R/V *Seward Johnson,* the ship that would be carrying us all to Cuba, felt like the makings of a three-ring circus. In ring one was the ship and sub crew, a well-known, highly dependable group. In ring two we had the science team, which was top-heavy with senior-level scientists. The absence of technicians and graduate students, who are often critical members of a research team, was evidence that the focus was going to be on form over function. There were four Harbor Branch scientists, including the chief scientist for the expedition, Grant Gilmore, plus four non-HBOI scientists—two from the United States, who were riding down with us, and two from Cuba, who would meet us there. And in ring three we had the film crew. These folks fell into two camps: the above-water team, led by coproducer Jimmy Lipscomb, a tall, thin, thoughtful documentarian, and the underwater team, led by Al Giddings, a bull of a man in both stature and personality, who'd been involved in underwater cinematography for a long string of high-profile films, including *The Deep, For Your Eyes Only, Never Cry Wolf, The Abyss,* and *Titanic.*

As the meeting progressed, further evidence that science was going to be taking a back seat to staged adventure emerged as the discussion focused on potential wrecks we might dive on along the southern coast. Wreck diving is fun, sure, but it had no scientific value for any of the scientists on this expedition.

Also on board was a corporate producer from Discovery Channel, a nervous little man who, I was relieved to learn, would not be joining us for the expedition. The opposite of a seasoned explorer, he was stressing out about *everything,* including the working title for the production: *Cuba: Forbidden Waters.* Fearful of rabid anti-Castro sentiments from powerful and highly vocal Cuban exiles and desperate to avoid political controversy of any kind, he was advocating for the far more white-bread title *Cuba: Enchanted Waters.*

In this same vein, he was also promoting the bizarre notion that there should be no mention of Castro. Besides the fact that that would be like describing a zebra without mentioning its stripes, it was clearly at odds with the whole concept that Giddings and Lipscomb had used to conceive of and promote the documentary. Their focus on the opportunity to explore a place that had previously been verboten clearly didn't fit corporate's need for a controversy-free production.

Effectively sharing science with the public hinges on being able to spin a good tale. Clearly, we had the essential elements: fantastic underwater visuals captured by Giddings combined with Lipscomb's erudite storytelling, revealing the thrill of exploring a heretofore inaccessible frontier. Lipscomb and Giddings had even taken the unusual step of including a political science scholar on the team. Richard Fagen, a professor of Latin American studies, recently retired from Stanford University, was introduced as someone who would be briefing us on the history and politics of places we would be visiting along the way.

A gifted storyteller in his own right, Fagen put the expedition on a whole new level for me. There's an old joke among oceanographers: "Become an oceanographer and see the ocean," the point being that although we often get to travel to some pretty exotic places for our research, it's not as glamorous as it sounds, because from the deck of an oceanographic vessel, one patch of ocean looks pretty much like every other. However, in this case we were going to be allowed to go ashore at frequent intervals, in an attempt to link the onshore politics with the offshore ecosystems.

I started to appreciate the significance of that linkage with our first stop, Santiago Harbor. Located on the southeast coast of the island of Cuba, Santiago was once a bustling seaport, but since the fall of the Soviet Union and the loss of trade with Russia, it had become a mere shadow of its former commercial self.

As we waited below the ramparts of the old Morro Castle for the harbor pilot who would guide us safely into port, we saw no other ships entering or leaving. In fact, the only vessels I spotted under power that day were the pilot's boat and a small gunboat that served as our escort. Otherwise, except for a few human-powered rowboats and sculls, all vessels were tied up or in dry dock. As a consequence, unlike in any port I had ever seen, there was no oily sheen on the water and no smell of diesel in the air, but rather a rich floral scent wafting off the land, which was largely undeveloped and covered with dense vegetation, in many places blanketing steep slopes right down to the water's edge.

As we approached the dock, we began to see people and traffic on the roads, but except for a rare bus, truck, or vintage car it was mostly bikes and a few horse-drawn carriages. As Fagen had explained in the briefing he gave the day before, the general lack of powered transportation was one of the hardships the Cubans had to contend with during this "special period," as he termed it, since supplies of oil had largely dried up with the dissolution of the Soviet Union.

Once we were tied up at the dock, we were told that, before we could do anything else, we must lower our U.S. flag so that it would be below the Cuban flag we were also flying. Although that didn't seem to bode well, all our subsequent interactions with the officials who came aboard to check our passports and stamp our papers went smoothly, and I was surprised to learn that we were free to go ashore and sightsee at will. We split up into small groups, with no clear plan beyond a desire to explore.

The small cohort I was with strolled around the waterfront and eventually wandered into a Cuban cigar factory. There was no machinery. It was just a large room, redolent with the sweet, woody smell of unburnt tobacco. Men and women sat at desklike workstations with little mounds of tobacco leaves that they were hand-rolling into cigars. A calico cat roamed among the stations, rubbing up against the workers' legs. We were told that lectors took turns reading novels aloud in order to keep the workers' minds occupied as they carried out their monotonous handiwork.

From there we strolled up to the nearby cathedral and central square, noting along the way how clean and uncrowded the city appeared. Although some of the buildings were crumbling, there were many splendid structures, with arches, columns, cornices, stained glass, and elaborate ironwork. There was also laundry hanging from many of the windows, balconies, and railings.

Eventually we ended up in a beautiful old hotel bar (Hotel Casa Granda) on the main square, where our crew had long since congregated. I saw Hector, a member of the sub crew who was fluent in Spanish, speaking with a number of the locals, and afterwards I asked him what he had learned. He said they were friendly, but very, very tired of the deprivation. They spoke of the animosities between our governments and made a clear distinction between their dislike of the American government and their fondness for the American people. Hector described how one of them said it had been thirty-six years since they had seen a ship flying an American flag in their harbor and they wanted to know if there would be more soon.

The contrast between what we saw on land and what we were observing underwater had the makings of a compelling story, if only Lipscomb and Giddings were allowed to tell it. The lack of industry, limited powered transportation, and a paucity of development along the shoreline made for some of the clearest, cleanest coastal waters I've ever seen. Nearshore seamounts, which are like undersea islands, were rich with healthy coral and sponges, but they were also devoid of big fish and covered with monofilament lines and anchors. Cubans' need to put protein on the table made fishing essential, and although lip service was given to maintaining sustainable fisheries, the need was too great and the protections too few. It was a case study in the tragedy of the commons.

There were nonetheless abundant small tropical fish that made the fish biologists on our team gleeful. Soon after we left Santiago, one of the submersible dives returned with a little orange fish. It was a bottom dweller called *Chaunax,* a member of the sea toad family. Placed in the aquarium in the wet lab, it sat obligingly on the black gravel (chosen for maximum contrast), and Giddings went to work with his high-resolution cameras getting the perfect artistic shot while Lipscomb and his crew filmed Giddings filming the fish and the fish biologists gathered around acting as talking heads, waxing poetic about the hapless creature. There were so many people crammed into the space around the tank that there wasn't room to move. Not wishing to add to the mayhem, I eased out the back of the lab and, in passing, overheard one of the crew members mumbling, "What's the big deal? It looks like the goldfish I had when I was a kid."

That fish encapsulated several of the storytelling challenges that documentarians face, which often boil down to striking the right balance between telling a good story and telling a true one. The first order of business for any successful nature documentary is vivid imagery. The adage that a picture is worth a thousand words has extra import in the deep sea, a realm so filled with bizarre and seemingly alien life-forms that they often defy imagination. These creatures not only are often fragile and singularly inaccessible but they live in the dark, which presents added difficulties if the goal is to film their natural behaviors.

It is hard to film animals unobtrusively, so that our presence doesn't influence their behavior. Lighting is obviously a big problem when dealing with nocturnal or deep-sea animals. On land, infrared-sensitive cameras in combination with infrared lights that are invisible to most animals provide a solution, but in the deep sea this isn't an option, because infrared light is absorbed so completely by water that it is rendered useless. Water also makes it impossible to use a telephoto lens to film animals from any great distance, as is so often done to observe skittish animals on land. The light-scattering properties of water make it necessary to be extremely close to your subject to get a clear shot. There are a few animals that will hold still long enough to permit a good close-up, but most won't, and trying to line up a high-resolution camera attached to a thirteen-ton submersible is no mean feat, especially if the animal is moving even the slightest bit. When possible, it's far better to capture the creature and hold it in a confined space.

Filming bottom-dwelling fish that are used to sitting on the seafloor, like *Chaunax*, is no great struggle, but taking footage of a captured midwater animal is much, much trickier. Creatures that have never encountered a surface at any time in their lives are prone to freaking out when they run into the side of the aquarium. The usual outcome involves the subject lying on its side or back on the bottom of the tank, in a decidedly unnatural pose. One way around this is to not even pretend that the close-up is happening in situ and instead show the aquarium and the scientists observing the animal, as Giddings and Lipscomb were doing. It's the most honest approach, but it's also kind of boring. Once or twice for a special animal is fine, but if that's your entire story, you've lost your audience.

The other approach is to use various degrees of fakery. One trick is to film close-up shots of captured specimens in a tank and pretend they're in their natural habitat. I saw a brilliant example of this for an Emmy Award–winning National Geographic production I was involved with, called *Ocean Drifters*. In order to show close-ups of a baby turtle encountering various forms of life in a floating sargassum seaweed mat, the crew used a 23,000-gallon tank with an optical-grade window. A wave generator was installed, then sargassum and animals collected from the Gulf Stream were introduced into the tank. The result was a unique floating world revealed through close-ups of a seahorse, a sea slug, a crab, and a sargassum fish, each perfectly camouflaged to match the sargassum they were floating among. It was amazing camera wizardry and a great way to share a wonderful bit of ocean life with the audience.

While the stated goal of natural history documentaries may be to educate the audience about the natural world, the commercial goal that makes it all possible relies on being able to entertain and engage the audience. That means the film needs to be more than just a string of natural history facts. You *must* tell a story. The baby turtle provided a clever way to narrate a tale about ocean drifters, the turtle serving as the connecting thread to introduce different open-ocean environments and inhabitants.

To build tension and add drama, they also included a scene where they introduced predators into the tank in the form of dolphinfish (a.k.a. mahi-mahi). Predators stalking prey is classic natural history theater, but it often requires some degree of fakery. In the case of *Ocean Drifters*, the dolphinfish were filmed swimming fast and attacking something near the edge of the weed line, and those scenes were intercut with close-ups of the baby turtle climbing up into the sargassum and tucking its hind flippers under its shell—in other words, looking scared. It was very endearing and made you root for the turtle, but those close-ups were filmed when the fish weren't in the tank. This is considered a no-no according to BBC guidelines for natural history documentaries, which specifically state that it is "normally unacceptable to . . . inter-cut shots and sequences to suggest they were happening at the same time, if the resulting juxtaposition of material leads to a distorted and misleading impression of events."

The BBC Natural History Unit has one of the best reputations for honesty of any of the major documentary producers, but even they have crossed the line from time to time. Their *Frozen Planet* documentary series contained some wonderful shots of polar bear cubs in an ice den, which it turned out were actually filmed in a German zoo's polar bear enclosure. There's no question that the viewers were given the misleading impression that the scene had been filmed in the wild, but, in point of fact, BBC *did* have a video posted on its website detailing how they got the shot. Given that this is a scene that could probably never be filmed in the wild without endangering the lives of the filmographers and/or the cubs, I think some leniency is in order. There are two ways that natural history documentaries can lose their audience: by being boring and by being dishonest. Unfortunately, data shows the former is far more deadly for programming than the latter, which means that producers are under extreme pressure to get the most dramatic shots possible.

Television and movies often give their audience a distorted sense of reality. How many people who watched the movie *Speed* actually believed a bus going seventy miles per hour can sail over a fifty-foot gap in the freeway? Some distortions are so pervasive that they are practically an industry standard, like visible laser beams,* bright red bloodstains that are supposedly days old,† audible explosions in the

---

* Lasers project narrow beams of light, and light must reflect off something to be seen, so unless you're in a smoke-filled dance club, the transmitted light is invisible until it hits its target—as anyone who has ever used a laser pointer well knows.

† Older blood turns brown from oxidation—as anyone who does laundry for "active" kids knows.

vacuum of space,* and the savage nature of sharks. In the latter case, the incredible commercial success of *Jaws*, first as a book and then as a film, played a major role in the vilification of sharks, culminating in Discovery Channel's Shark Week, which has demonstrated many extreme examples of distorting events. Titles such as "Great White Serial Killer Lives," "Australia's Deadliest Shark Attacks," *Tiger Shark Terror*, and *Voodoo Shark* give a chilling impression of these magnificent animals, belying the fact that while sharks are generally responsible for about six human fatalities worldwide each year—often as a result of mistaken identity—humans are responsible for the deaths of approximately *one hundred million* sharks per year! Yet attempts to institute policies for the protection of sharks are often met with lukewarm enthusiasm from people who have bought into the hype that sharks are bloodthirsty killers.

The most extreme example of Discovery Channel crossing the line is its fake documentary *Megalodon: The Monster Shark Lives*, which it used to kick off Shark Week 2013. It was billed as "the rediscovery of a giant prehistoric predator thought to have been extinct for more than thirty-five million years." The whole mess deserves a pants-on-fire rating of 11, since they used actors to portray scientists, fabricated computer-generated imagery (CGI) that they claimed was video evidence, and showed fictional scenes overlaid with the declaration that "what you are witnessing are the actual events as they unfolded." Only in the final seconds of the "documentary" did a series of mealymouthed disclaimers appear, flashing across the screen at a rate that would have left an Evelyn Wood speed reader spluttering:

---

* Sound waves require a transmissive medium like air or water in order to stimulate your eardrum—as anyone who has survived time in a vacuum can tell you. This, of course, is sarcasm, but not because humans can't survive a vacuum (you have fifteen seconds before you pass out) but because who do you know that's ever actually been in a vacuum?

None of the institutions or agencies that appear in the film are affiliated with it in any way, nor have approved its contents. Though certain events and characters in this film have been dramatized, sightings of "Submarine" continue to this day. Megalodon was a real shark. Legends of giant sharks persist all over the world. There is still debate about what they might be.

Although there was an outcry when the ruse was revealed, the show broke records, drawing 4.8 million viewers, which made it an outrageous success as far as the producers were concerned.

Another egregious example of fakery was *Mermaids: The Body Found*, which aired on Animal Planet. A work of complete science fiction, it purported to uncover a plot by the government to conceal evidence of mermaids. It was done in documentary style, with fake "reenactments," lots of CGI, actors playing scientists, and a too brief disclaimer at the end that most viewers missed. They even took the stunt so far as to have a fake web page pop up if you tried to look up the supposed former National Oceanic and Atmospheric Administration scientist whistleblower. The page displayed a Department of Justice logo and a Homeland Security Investigations special agent's badge shown above the declaration "This domain name has been seized by DOJ–Homeland Security Investigations, pursuant to a seizure warrant issued by a United States District Court under the authority of 18 U.S.C. §§ 286, §§ 287 and §§ 371." This was a total fabrication posted by Animal Planet's parent company, Discovery Communications, and clear evidence of just how far over the line they were willing to go in a craven grab for ratings.

As with *Megalodon,* there was backlash, but given that this show and its sequel, *Mermaids: The New Evidence,* were the most watched programs in Animal Planet's history, Discovery Communications— which also operates Discovery Channel, TLC, and more than 150 other worldwide cable TV networks—was unmoved. They set out to fool their audience and, unfortunately, in far too many cases they succeeded. So many people believed that the whistleblower NOAA scientist, played by actor Andre Weideman, was real and the victim of a plot by NOAA to cover up evidence of mermaids that NOAA felt compelled to take the very unusual and slightly surreal step of posting a statement on its website that "no evidence of aquatic humanoids has ever been found." The whole thing would be funny except that, like Shark Week, it resulted in chilling collateral damage: the undermining of the public's trust in science. The impact is best encapsulated by this statement by a fifth-grade schoolteacher who watched the program: "If NOAA is lying to us about the existence of mermaids then they're definitely lying to us about climate change."*

How can the public be expected to make sound decisions about important issues like climate change and shark preservation when a major cable network like Discovery Communications, which actually bills itself as "the world's #1 nonfiction media company," is peddling programming like *Mermaids* and *Megalodon*? It is a dangerous gambit—one that not only negatively impacts important policy decisions but has also tarnished Discovery's reputation and led many scientists to refuse to work with them. It's a downward spiral with grim portents for life in a post-truth world, but fortunately, someone in corporate who had actually heard and understood the children's story "The Boy Who Cried Wolf" declared that, beginning in 2015, Discovery was turning over a new leaf and would produce no more fake documentaries. It's a start, but once you've betrayed trust, it's difficult to earn it back. I hope they do, because Discovery Communications has a great big megaphone that could go a long way toward raising the level of the public's understanding of science.

* Many scientists spoke out about the damage done by the *Mermaids* programs. One of these was marine biologist Andrew David Thaler, who wrote an article for *Slate* ("The Politics of Fake Documentaries," August 31, 2016) in which he related this quote—which made my toes curl.

* * *

DURING THE CUBA expedition, we encountered a classic example of the difficulties associated with packaging science for public consumption when filmmakers faced an unexpected setback: Unlike on most deep-sea expeditions, during which you can pretty much count on encountering some new animal or behavior, we were seeing little to get excited about. The JSL submersible depth rating had been downgraded from three thousand to two thousand feet just before the expedition. Because the water was so clear, this meant we weren't able to get below the edge of darkness during daytime dives, and the only animals we were seeing were small fish hiding among the rocks along the island's steep underwater slopes and small, transparent creatures in the midwater. As a result, in what I suspect was a slightly desperate attempt to insert story where there was none, Lipscomb started asking each of us to describe our worst experience in a submersible—while on camera. He also wanted us to discuss safety issues and explain what had gone wrong with the sub, that it needed its depth limit adjusted. I demurred and later found out that the other scientists had also. Even though I liked Lipscomb, I simply didn't trust that anything I said wouldn't be sensationalized.

The Johnson-Sea-Links had many noteworthy distinctions that set them apart as research tools. Unfortunately, in addition to their numerous achievements, they were known for "the accident." Being involved in the only fatalities ever to occur with a research submersible was not a point of pride.

The story was legendary to divers. It was Father's Day, June 17, 1973, two years after Edwin Link invented and launched the first Johnson-Sea-Link submersible. Link and his team, which included his thirty-one-year-old son, Clayton Link, were on an expedition in Key West. It was the sub's 130th descent, and there was, in retrospect, possibly an overly relaxed attitude to the dangers involved. It was supposed to be a short and relatively shallow dive to a sunken ship—a retired Navy destroyer that had been purposely sunk to create an artificial reef. Just fifteen miles offshore, in only 360 feet of water, the mission was seemingly simple—to retrieve a fish trap from the deck of the destroyer. The sub had a crew of four. Up front in the observation sphere was the pilot, Archibald "Jock" Menzies, with about a hundred JSL dives under his belt, and sitting next to him was fish biologist Robert Meek. In the separate rear dive chamber were Clayton Link and submersible crew member Albert "Smoky" Stover.

As accidents go, it didn't seem all that dramatic—at least at the start. As they moved in on the trap, a current pushed the sub into a snarl of cables on the wreck from which it couldn't break free. As one attempt after another to liberate the entangled sub failed, the carbon dioxide levels in both chambers were climbing. In those days, the chemical used to scrub the air was Baralyme, which did not work well at cold temperatures. Because the surrounding water temperature was a chilly forty-five degrees Fahrenheit, both chambers were cooling, but the rear aluminum chamber much more so than the acrylic sphere. This was making things uncomfortable for Link and Stover, who had not bothered to bring any warm clothing for what was supposed to be a short dive; they were wearing only T-shirts and shorts.

Besides the penetrating cold, they were also suffering from the pounding headaches and labored breathing characteristic of CO_2 poisoning. When they eventually realized that the low temperature was why the carbon dioxide levels were getting so much higher for them than for Menzies and Meek, they tried rubbing the Baralyme on their bodies to raise the chemical's temperature. It was no use. Approximately twenty hours after the sub became trapped, the pilot, Menzies, called up to the surface to report that he could hear Link and Stover suffering convulsions in the dive chamber. By the time the sub was retrieved, more than eleven hours later, using a grappling hook deployed from a commercial salvage vessel, it was too late. Link and Stover had died from carbon dioxide poisoning.

For Edwin Link to lose his son in the submersible he had designed, and on Father's Day no less, seemed the cruelest imaginable
twist of fate. Since first learning the particulars of the accident, I had
tried not to imagine what it must have been like for Edwin up on the
deck of his ship, desperately trying to make all the right moves to
free his trapped son and crewmates below. I remember thinking that
at least he was spared the excruciating agony of hearing what Menzies heard. Which is why my worst moment in a submersible was
not the time I found seawater streaming in through the inlet valve
on Deep Rover. It was the time water came into the JSL while I was
in the acrylic sphere and my husband was in the rear dive chamber,
where the water was coming in.

How did my husband come to be diving in a submersible with
me? Well, after David graduated from Brooks Institute of Photography, he worked for a while in video production and became interested in the engineering side of the industry. I was still at UCSB, so
he decided to go back to school and get a master's degree in physics
instrumentation—ironically, the same program I had been discour-
aged from completing. He graduated with the equivalent of a degree
in computer engineering, which made him highly employable—so
much so that several months after I started at Harbor Branch Oceanographic Institute in its marine science division, he was hired on in
its engineering division, which allowed us the opportunity to work
together on certain projects.

In 1991, I was chief scientist on a JSL expedition to the Bahamas
and I had talked David into coming along by promising him his
first-ever dive in a submersible. I delivered on that promise on the
opening day of the expedition. Pilot Phil Santos and I sat up front in
the acrylic sphere, while David and crewman Kruno Rehak rode in
the dive chamber. It was a beautiful warm Bahamian day in February, but since we were going to three thousand feet, where the temperature can drop below forty degrees Fahrenheit, I had cautioned
David to bring a jacket.

I was elated to share the thrill of exploring the deep sea with him and anxious that he enjoy it. With that in mind, I had also warned him about Kruno's warped sense of humor. On one expedition, Kruno had startled one of my British colleagues when, on his first dive in the dive chamber, after giving the safety briefing, Kruno had flipped open a switchblade and said, "Know what this is? An oxygen doubler." Fortunately, David seemed well liked by the sub crew and was not subjected to any of their usual hazing.*

David and I couldn't see each other, but we could talk through our headsets. The protocol was for the science observer in the back to take notes for the scientist in the front, writing down key sightings, along with depth, time, and temperatures at which organisms were collected. As we descended, David was looking out the small port-hole on the starboard side of the dive chamber while I enjoyed the far more expansive view afforded by the acrylic sphere.

I took on the role of tour guide, pointing out the dramatic color changes, from crystalline turquoise at the surface through the grad-ual shifts to muted azure and then smoky navy and charcoal Prus-sian blues and finally fading to gray and black, interrupted by streaks and flashes of electric blue as we began to see bioluminescence at around twelve hundred feet. After watching the bioluminescence for a while, Phil turned on the lights so that we could spot animals in the water column, but I was disappointed to see little life. As we ap-proached the bottom, I was on the lookout for the pelagic sea cu-cumber *Enypniastes*, which we wanted to both film and collect, when Kruno delivered the alarming news that the seawater inlet valve[†] was leaking.

---

*They would sometimes caution first-timers to drink lots of coffee before the dive to help them stay warm. Then, when the victim arrived on deck after a typical three-and-a-half-hour dive with bladder bursting, as an added torture they would make sure the nearest head was occupied.

†The JSLs were designed for scientific research, which sometimes involved col-lecting water samples.

Kruno's revelation of the leak triggered a blast of adrenaline through my body. Stories of the infamous "accident" came rushing back to me, and I was suddenly absolutely certain that my husband was going to die a horrible death in the dive chamber while I could do nothing but listen. The fear was so intense that, for the first and only time in my life, I experienced synesthesia—my senses momentarily interlinked and I simultaneously felt and saw the fear as a blinding blue light.

Kruno said it was a drip, not a flood, but it was increasing—not a good sign. "Phil, we need to ascend *now*," I barked, but he shook his head, saying, "That could take too long." Once a leak starts, it can accelerate rapidly. The metal high-pressure tubing for the inlet passed through the bottom of the acrylic sphere, and there was a backup valve that could be shut off, if only he could reach it. Straightaway, Phil began ripping out all the electronic gear between our two seats, piling the video recorders and camera controllers in a teetering stack on my lap as he attempted to clear a path to the valve. Would it turn off or behave like the one I had dealt with in Deep Rover and refuse to budge?

When the last piece of gear was cleared away, Phil stretched down as far as he could, past the various flanges and tube fittings, to reach the valve. When he got a grip, he cranked on it hard. The handle turned, and almost immediately Kruno reported that the water had stopped. I have no sense of how long it was between when Kruno first reported the leak and when he declared it had stopped, but it unquestionably represented my worst stretch ever in a submersible. Phil didn't say much about it at the time, but as we reassembled the electronics in order to continue the dive, I noted that the backs of his hands and forearms were dotted with bleeding scratches from his attempts to reach the valve. Evidently it had been a bad moment for him, too.

I wouldn't know how bad until six years later, while watching the broadcast of our Cuba trip. When Lipscomb asked him about his worst experience in a submersible, he described that leak. Phil admitted, "No matter how well prepared you think you are, the first reaction is panic . . . *That could have been my last dive.* And that's something you hate to think about, but you have to." Only in hearing Phil's account on the documentary did it occur to me that Phil and I might have died, too, albeit long after David and Kruno.

ALTHOUGH THE CHANCES are infinitesimal,[*] entrapment is the most likely cause of death in a submersible. Catastrophic implosion of the acrylic sphere is far less likely. Still, concern about that possibility had caused the JSL's depth limit to be downgraded from three thousand feet to two thousand feet.

Making a submersible out of acrylic was a novel concept when Ed Link took it on in 1971. In 1970, the Navy had just completed construction of the first transparent-hull submersible, the NEMO (Naval Experimental Manned Observatory), which had a depth limit of six hundred feet. The JSL 1, which Link launched in 1971, greatly pushed the depth envelope by increasing the limit to two thousand feet. In fact, the engineering specs indicated that the hull design could withstand even deeper dives, which is why, in an effort to push the frontiers of deep-sea exploration, HBOI eventually started diving the subs to 2,640 feet. However, when cracks began to appear in the acrylic, new spheres were constructed that increased the hull-wall thickness from 4 inches to 5.25 inches. These came online before I arrived at HBOI, in 1989, so when I started using them, the subs were routinely making dives to three thousand feet.

Both of HBOI's JSLs were highly dependable research platforms, with one little idiosyncrasy in JSL 1: After a dive to three thousand feet, during the ascent, just as you were nearing the surface at around

---

[*] Of the hundreds of thousands of dives made in research submersibles, there have been only two fatalities. It is far more dangerous to drive in Boston than to dive in a submersible.

two hundred feet, the sub would sometimes produce a loud bang. The first time I heard it, the pilot had warned me we might hear a bang, but HOLY HEART FAILURE BATMAN! It was so loud I practically levitated. I later overheard the sub crew debating the possible cause. They were uncertain, which was not reassuring. In the end, it was determined to be a design flaw related to the compressibility of acrylic.

The fact that the acrylic sphere compressed under pressure meant that its diameter actually decreased by as much as half an inch at depth. I have never tried this, but I was told that if you stretched a string taut from one side of the sphere to the other, once you reached three thousand feet, it would be hanging loose—only to retighten when you returned to the surface. To deal with this compression, the aluminum hatch through which we entered at the top had beveled sides that sat in a beveled hole in the acrylic, surrounded by a nylon gasket ring that was supposed to allow the hatch to slide outward as the sphere compressed, and inward as it expanded.

The trouble was that it wasn't sliding during the ascent, but rather sticking and then letting go all at once, with two unwanted results: the loud bang as it released, and the development of tiny shear cracks at the interface. By 1997, those cracks had grown to the size of quarters. Reducing the depth limit to two thousand feet eradicated the bang and the cracking, but the utility of the subs for deep-sea exploration was severely impeded. Eventually, the problem was solved with a new sphere designed with different interface angles, a thicker gasket, and a new lubricant. But all that happened *after* the Cuba mission.

THE CUBA EXPEDITION was not shaping up the way Lipscomb and Giddings had planned. They were getting good stuff, but, thanks to the depth limitations, most of it was shot during shallow-water dives or while ashore. Nothing terribly exciting had happened or been discovered, and the rarity of the locale was not shining through.

With the expedition drawing to a close, the New Year dawned with ten-to-twelve-foot seas and a stiff headwind as we steamed toward Havana Harbor, our final stop before making our way home. There were two big things that Giddings was hoping might happen once we got there. The first was for the sub to dive in what he called the "boneyard," a six-hundred-foot-deep gully at the entrance to Havana Harbor, a place that had presumably collected a half century's worth of relics and treasures from the trading vessels and treasure ships that plied these waters. The second was a visit from Castro.

The reason we had been able to get clearance to dive these waters was primarily because Al Giddings had a personal relationship with Fidel Castro—one that was based on their shared fondness for scuba diving and ocean exploration. They had bonded over their common love of the ocean when Giddings visited Cuba on filming expeditions in the late seventies and again in the early eighties, and from the outset of our expedition, Giddings had been talking about possibly taking Castro on a submersible dive.

Lipscomb was made practically apoplectic by the notion, and it had become a major point of friction between the two. The very idea of carrying the seventy-one-year-old dictator into the deep sea was problematic on many levels, but the worry Lipscomb raised that I found most compelling was "What if he had a heart attack while he was in the sub? There's no way they would believe it was an accident. They'd never let us leave the country. For Christ's sake, they'd kill us!"

Both the boneyard dive and Castro's visit to the ship happened on the same day: January 2, 1998. The dive in the morning was a total bust, because as the sub descended the sloping wall of the gully, it got caught in a current and pushed into a snarl of heavy-duty undersea cables almost halfway down, at around 320 feet. For the sub crew, this was too reminiscent of the circumstances of "the accident," and the decision to abort was immediate and nonnegotiable.

So Castro's visit was Giddings and Lipscomb's best hope for a grand finale. We still weren't sure it was really going to happen. It was supposed to be a secret, for security reasons, but when we were all advised to remain on board that evening instead of going ashore to the wonderful outdoor cafés with their live music, we guessed the reason. Just after sunset, a gunboat quietly positioned itself off our port side, followed shortly after by three black Mercedeses speeding up to the end of the gangplank on our starboard side. With no fanfare, presidential guards in olive-green fatigues, packing automatic weapons, along with bodyguards in civilian garb with no visible weapons* spilled out as El Comandante emerged from the middle car. Dressed in his standard fatigues, his hair and beard showing very little gray, he looked a lot younger than seventy-one.

Giddings served as official greeter and master of ceremonies, ushering Castro aboard and making introductions. There was practically a scrum on the fantail as everyone clustered around, jockeying for position, while Giddings and the chief submersible pilot took turns describing the submersible and its many capabilities. A petite woman in a denim jacket served as simultaneous translator, a job she excelled at, even to the point of repeating the mannerisms and gestures of whomever she was translating at the moment. Much of the conversation was driven by Castro's questions, of which there were many. He seemed intensely interested in all aspects of our operation.

Everyone, not just the film crews, wielded cameras. We all wanted to document this historic moment. I was holding the camcorder that I used in the sub, but most of my shots were from the back of the scrum. In hopes of getting some good close-ups, Tammy and I positioned ourselves in a doorway along the narrow hallway leading forward to the galley, figuring he would necessarily walk by that spot during the tour of the ship. He did come that way, and I was filming, but instead of just walking by, he stopped to talk.

---

*We later learned they each had a gun hidden under their untucked shirts— a fact that came to light during the ship's tour, when someone made a too sudden move to open a drawer. The drawer contained a T-shirt intended as a gift, not a weapon, but although the security appeared relaxed, these guys weren't.

I quickly dropped the camera to my side, freeing up my right hand to shake his. I was surprised at the softness of his grip. It seemed at odds with the persona. I was even more taken aback by the scientific depth of his questions to us both. He wasn't just showing off what he knew—which was a great deal—he was trying to learn *new* things. In answer to his question "What do you study?" I had begun describing a world without sunlight, where animals still have eyes to see the light that they themselves make, when he broke in with what seemed to be a non sequitur: "Why don't they freeze?" But further questions revealed that what he wanted to know was, if there was no sunlight to warm the depths and the temperatures were so close to freezing, why didn't the fish freeze? It's a smart question, with a complex answer having to do with the Earth's hot core and thin oceanic crust, ocean circulation patterns, and how salt lowers the freezing point of water. He took that in, and then asked if there was evidence of El Niño and climate change in the deep sea. He spoke at length about climate change and its implications for island nations like the Maldives and about how agriculture and fisheries will be impacted worldwide. Finally, he talked about the decimation of shark populations because of finning. It was a wholly unexpected dialogue on many levels, and I was pleased to discover later that, because I had left the camera running, I had the whole thing on tape, along with what may be the most unusual footage of Fidel Castro ever shot. The way I was holding the camera, it was tilting straight up, affording a detailed view of the inside of the dictator's nose.

Later during his visit, I scored another coup when I happened to be standing in the galley as Castro was coming down from the bridge and he stepped out of the procession to ask our cook about how he prepared lobster. I was filming as he then proceeded to give a lecture on how you mustn't ruin lobster the way the French do. "You can't even taste the lobster in lobster thermidor. It's like mixing water with wine! Instead you just cut it open like a butterfly, add a little butter and onion, and cook it eleven minutes." Lobster à la Comandante.

Castro touched on so many subjects during his time on board the R/V *Seward Johnson*, most of it on camera, that Lipscomb and Giddings had a plethora of material they could use to flesh out their story. They had superb footage, for example, from one of the island's spiny lobster fisheries that they could now intercut with Castro describing the life history of spiny lobster and the importance of maintaining a sustainable fishery. There was the backstory of how the lobsters were fresh-frozen and shipped to Europe, creating one of the country's most valuable exports, but the Cuban people didn't get to eat any of that harvest. And then there was the irony, or humor (depending on how they wanted to play it), of El Comandante providing recipes for food his own people weren't allowed to eat. The point was that Castro's visit offered the basis for some great storytelling. The question was, would they be allowed to use any of it?

The answer turned out to be no. Corporate was just too concerned about potential backlash from the anti-Castro community and they chose to play it safe. The two-hour documentary, which was narrated by Martin Sheen, made no mention of Castro. The title was changed from *Cuba: Forbidden Waters* to *Cuba: Forbidden Depths,* with the explanation that "forbidden" referred not to the political restrictions but to the "treacherous depths" we were exploring. This was especially ridiculous given that we were diving to only two-thirds of our usual depth limit.

In the end, no one was terribly happy with the outcome. Lipscomb was so dissatisfied that he never even watched the television broadcast. Corporate was unhappy because the documentary got poor ratings. And the scientists were frustrated because so little real science was accomplished. It was a very unusual trip, though, and when it was over I felt privileged to have been able to explore and experience Cuba from a unique perspective. But when I got home, I vowed to David that if I was ever given the opportunity for another television-sponsored expedition, I would *never* accept. It was a vow I would eventually break.

Chapter 10

PLAN B

I typed in the command to retrieve the camera images. Nothing happened. I typed it in again. Still nothing. I felt my jaw clench but fought to keep my face from conveying the depth of my dismay. The inevitability of this moment seemed preordained. Murphy's law is immutable: *If anything can go wrong—it will.* And so is its corollary: *If there is a worst time for something to go wrong, it will happen then,* which in this case was right now, on national television.

This was exactly the scenario I had envisioned when filmmaker David Clark contacted me in 2003 about the possibility of his coming along on this mission. An Emmy Award–winning independent filmmaker with a good reputation, Clark was someone with whom I normally would have been happy to work, and since he would be coming along as my guest, it wouldn't be like the Cuba trip; I would be calling the shots. But this mission was special.

This was going to be the first field test of an instrument that I had been struggling to get funded for years. The possibility for failure is always high the first time you test a new instrument in the field, and I was less than thrilled at the notion of having such a failure made so very public. On the other hand, Clark wanted to focus the documentary on the importance of engineering in ocean exploration, which is something I feel strongly about. It was a fraught moment, but one that presented an opportunity to communicate crucial information to a large audience.

We must do a better job of helping people understand what it means to live on an ocean planet. More specifically: what it means to live on a few little dry islands surrounded by a vast watery world that we know surprisingly little about. We are overrunning the Earth with our population—now about to blow past eight billion. To feed our exploding numbers, we are intensively farming the land and stripping the ocean of its biomass while creating all manner of wastes, which spill off the land and into the water, overwhelming the complex life-support machinery that maintains us. This is not sustainable, and it is definitely not smart.

How is it possible, in this day and age, that we know so little about the workings of our own planet? The first step to knowing is exploration. So how much of the ocean have we actually explored? The answer you hear most often is 5 percent. The funny thing is, some will tell you that number is way too small, while others (me, for instance) will tell you that it's way too large. It all depends on what you mean by "exploration."

If a map is all you need to declare a region explored, then we can claim to have explored 100 percent of the ocean. However, that map was made from space, using satellites that scan the sea surface using radar. Radar doesn't penetrate seawater; it bounces off, providing very accurate measurements of sea surface heights but not much else. By taking lots of measurements and averaging out the bumpiness and oscillations produced by waves and tides, we reveal bottom features like undersea mountain ranges and trenches. Unfortunately, the resolution of that map is so poor that it is impossible to discern anything smaller than three miles across; anything less would be undetectable. Smaller features associated with seamounts, deep-sea vents, and the hills, ridges, canyons, and valleys that provide critical habitats for animals are unresolved.

We actually have more precise maps of the moon, Venus, and Mars.

Ships cruising along the surface using multibeam sonar projected in a narrow swath along the seafloor have produced higher-resolution maps of almost 30 percent of the ocean bottom. These maps provide a resolution of about a hundred yards, which is still not great. By comparison, the resolution of your house as seen on Google Earth is twenty-five inches.*

To see at higher resolutions in the ocean, we must physically pierce the watery veil. So if your definition of exploration involves actually visiting a place, we have explored less than a paltry 0.05 percent of the deep ocean! That would be like reconnoitering a mere three city blocks out of the entire island of Manhattan—*and* only at ground level, because even that tiny percentage completely ignores the staggering volume of living space *above* the seafloor, which, at an average depth of 2.3 miles, is equivalent to a building 1,207 stories tall.†

So what is the big hang-up that has kept us from exploring the vast majority of our planet? As the saying goes, it ain't rocket science. It's chronic underfunding. The fact is, we have never had anything akin to a moon shot or a NASA for the ocean.

When President Kennedy made his famous moon-shot speech in 1962, he painted a picture of space as a beckoning frontier. He said, "We set sail on this new sea because there is new knowledge to be gained, and new rights to be won." He focused on the need for the United States to establish a position of preeminence in order to protect ourselves "against the hostile misuse of space" and alluded to how we were losing the space race with the Soviet Union, with potentially scary consequences. His rhetoric went a long way toward

---

*I'm just guessing based on personal experience. It depends on where your house is. Resolution on Google Earth ranges between six inches and fifty feet.

†The Empire State Building is 102 stories.

selling the program, despite considerable pushback from those who felt it was a tremendous waste of money.

Kennedy's predecessor, Dwight D. Eisenhower, stated unequivocally that "to spend $40 billion to reach the moon is just nuts." Looking back now, most people would disagree with that assessment. Walking on the moon is widely characterized as an enormous human achievement and a shining example of American exceptionalism.

In the absence of any comparable geopolitical drivers or clearly defined goals, there has never been any serious long-term financial commitment to ocean exploration. In 2013, the United States budgeted $3.8 billion for space exploration, but just 0.6 percent as much—$23.7 million—for ocean exploration. Looked at from a different perspective, for the cost of one shuttle launch (with payload, about $1 billion) we could have financed two deep submersible dives a day (at $12,500 each) every day for 110 years. That's an enormous disparity that helps explain why we know so little about our ocean planet.

Scientific achievements have always been linked to technological advancements. But for innovations to occur, there must be a sustained source of funding, first to develop the technology and then to support its continued progress and applications. The recent remarkable technological achievement of being able to image a star nine billion light-years away was made possible by a very significant investment in the Hubble Space Telescope at a total long-term time investment of more than a third of a century and a total cost of well over $10 billion.

By way of contrast, a comparable investment for ocean exploration would be the deep submergence vehicle (DSV) *Alvin*. This little three-person deep-diving submersible, which was originally built for less than half a million dollars, is a great example of a technological advancement leading to incredible scientific discoveries. But in the absence of the kind of PR generated by the space program, those discoveries go largely unhonored and unsung. To most people, it's just a clunky-looking little sub with a funny name.

The idea of developing a submersible for science was not widely embraced when it was first proposed, since no one was sure what it might discover. It was Woods Hole scientist Allyn Vine, speaking at a national symposium in Washington, D.C., in 1956, who best articulated the need for a human presence in the ocean: "I believe firmly that a good instrument can measure almost anything better than a person can *if you know what you want to measure* . . . But people are so versatile, they can sense things to be done and can investigate problems. I find it difficult to imagine what kind of instrument should have been put on the *Beagle* instead of Charles Darwin" (emphasis mine). It was a speech with enough punch to sway opinion, so when the sub was finally built and launched, in 1964, it was christened *Alvin*, a contraction of Allyn Vine.

Alvin, which was paid for by the Office of Naval Research, was originally designed to dive to 8,010 feet* carrying a pilot and two passengers. Since then it has undergone several rebuilds and upgrades, the most recent in 2013, funded by the National Science Foundation (NSF). In all its various incarnations, the sub has had a long, illustrious career spanning more than half a century. Its long list of impressive achievements includes discovering the hydrothermal vents, recovering a lost hydrogen bomb off the southern coast of Spain, diving on the *Titanic*, photographing deepwater corals smothered in brown goo following the Deepwater Horizon oil spill, and a lengthy string of major scientific findings and breakthroughs, including several that have helped to radically transform our understanding of how our world works.

MY BELIEF IN the importance of submersibles grows out of direct experience. For the majority of my career, no camera system existed

---

* Currently rated to 14,760 feet.

that came close to the capabilities of the dark-adapted human eye for
seeing bioluminescence. As a result, I spent many hours in the deep
sea sitting in submersibles with the lights extinguished, seeing the
largest ecosystem on the planet as few others have. It gave me plenty
of time to think about what life must be like for the animals in this
realm, and I have often wondered how great our impact must be
when we enter their world with our screaming thrusters and blinding floodlights.

Those thoughts would resurrect childhood memories of summer
nights when the neighborhood kids would come together to play
hide-and-seek. We'd gather by the lamppost at the corner and then
disperse into the surrounding darkness of our suburban neighborhood to hide from whoever was "it." One of the best hiding places
was just outside the illumination halo of the streetlight in a neighbor's yard, lying flat on the ground where you could see the action
around home base but you wouldn't be seen. In the sub with the
lights on, I could imagine a spherical halo of animals around me,
lurking just outside the reach of my lights, playing their own games
of hide-and-seek. How could we ever hope to draw them in?

DESPITE MY PASSION for submersibles, I knew Allyn Vine was
right: If you know what you want to measure, then you can probably
develop a remote system to do it. In my case, I wanted to determine
what animals and behaviors I might see if I *wasn't* physically there,
scaring them away. A remote system was the logical solution.

What I needed was a battery-powered deep-sea camera that could
be left untended for extended periods. These existed, but they depended on white light. I wanted bioluminescence to be visible, which
meant turning out the lights. But at the same time, I wanted to be
able to see the animals. If I was going to be truly unobtrusive, I
needed an illumination system for the camera that the animals
couldn't see. I knew that red light had been tried by a number of investigators working from submersibles, but always with disappoint-

ing results. The light was absorbed over such short distances in water that it was essentially useless. My idea was to try to compensate for the extremely poor illumination that red light provides in water by using one of the super-intensified cameras I had been employing for recording bioluminescence stimulated by the SPLAT screen. If I got the illumination levels just right, I should be able to see both the animals and their bioluminescence. I had done the math and was convinced it should work. I even had what I thought was a cool name for it: the Eye-in-the-Sea. What I didn't have was the money to make it a reality.*

I will spare you the grant-writing blow-by-blow. The bottom line was that before any of the funding agencies would consider ponying up financial support, they wanted to know what, exactly, I would discover. I had no idea—that was the point! But I was convinced that there must be oodles of critters out there we didn't even know existed because we were scaring them away. Based on the proposal reviews I received, it was clear I was going to have to supply proof of concept, which meant field data demonstrating that it was possible to illuminate the animals in such a way that I could see them but they couldn't see me.

Deep-sea field studies are expensive. I was doing well managing to fund two or three major expeditions a year. But inevitably, the expense was such that the number of at-sea days that got funded was less than the time needed to do the research proposed—especially when you factored in days lost to bad weather and equipment failures. There was zero time remaining in my existing expeditions for screwing around with red lights.

It was 1994 when I wrote my first proposal for the Eye-in-the-Sea (EITS). Six years passed before I finally found a way to establish

---

* Here's a useful tip if you are ever suffering from insomnia: Call up a scientist and ask about their grant funding. You will be treated to a sleep-inducing, interminable tale of woe worthy of a Greek tragedy.

proof of concept, thanks to the Monterey Bay Aquarium Research Institute. MBARI (pronounced *em-BAR-ee*) is a world-class research institution that traces its origins to our 1985 Deep Rover expedition in the Monterey Canyon and, more specifically, to a series of conversations that grew out of that mission between Bruce Robison (the chief scientist on our Wasp and Deep Rover expeditions) and David Packard (of Hewlett-Packard fame and fortune).

Packard had financed the hugely successful Monterey Bay Aquarium, which opened in 1984 on the site of the last sardine cannery along Cannery Row. He was a strong proponent of science and had been envisioning a research program associated with the aquarium, but with Robi acting as his personal pied piper into the deep sea, aided by the glorious high-resolution videos we were bringing back from our Deep Rover dives, he started entertaining a greatly expanded vision. Why not an entire research institution? One that would take advantage of the unusual underwater topography of Monterey Bay, which put the head of a massive, biologically rich subsea canyon a mere stone's throw from shore?

Not surprisingly, Robi was hired on by the newly formed institution and had been working there since 1987, using MBARI's state-of-the-art remotely operated vehicles (ROVs) to study deep-sea life in the Monterey Canyon. He suggested that I apply for an adjunct appointment at MBARI, which had the amazingly generous policy of providing its adjuncts with free ship and ROV time to conduct research. I applied and was thrilled to be accepted. Finally, I had an opportunity to experiment with red lights. I didn't, as yet, have a battery-powered camera that I could leave on the bottom. So for my first attempt, in 2000, which was to be a two-day trial run, I planned to use their ROV *Ventana* to carry a bait box down to the bottom to attract animals that I would observe with my intensified camera, wired into the ROV, spotlit by alternating red and white light for illumination. I hoped to be able to prove to potential funding agencies that red illumination would work with this arrangement and, if possible, provide evidence of more animals seen under red compared to white light.

Several friends had warned me that MBARI's research vessel, the *Point Lobos,* had a reputation as a "vomit comet," but I figured that would be no problem for me. I had been going to sea for sixteen years and had been on countless similar vessels without effect. So when I first started feeling a bit off just a couple of hours into the mission, I chalked it up to jet lag, since I had just gotten off a plane from Florida.

It is said that there are five stages of seasickness. Stage 1 is denial. Stage 2 is nausea, which, by the time we reached the dive site, I was rapidly approaching. Stage 3 is feeding the fish. After that comes stage 4, when you're afraid you're going to die, followed by stage 5, when you're afraid you're not.* There were a couple of reasons why being seasick was out of the question. First, this was the beginning of a series of missions with this vessel and this team, so throwing chow was *not* the first impression I wanted to make. And second, Robi had come along to help train me on ROV ops, so I needed to focus.

When we reached the dive site, I went out on deck, where I breathed in great gulps of fresh air and tried to focus on all the activity associated with launching the ROV. The *Ventana* stood tall on the aft deck, with the dimensions and utility of a toolshed: eight feet high, six feet wide, eight feet long, and bristling with gear, including cameras, lights, manipulators, and all manner of samplers for collecting animals. The top third was constructed of molded syntactic foam painted tangerine orange, with the MBARI logo—a sinuous deep-sea gulper eel poised in the cleft of a jagged V, symbolizing the

---

*This is real. Over the years I have heard several credible accounts of people reaching stage 5 and having to be physically restrained to keep them from jumping overboard to end their misery.

Monterey Canyon—emblazoned in blue. The bottom two-thirds was a dense snarl of instrumentation and cables.

Launched off the starboard side of the ship, the 7,500-pound ROV was plucked off the deck by a stern-mounted crane that did a stiff-arm lift, popping it into the ocean in a matter of seconds. As soon as it was in the water, the crane operator released its grip and the ROV pilot standing on deck took control by using a belly pack fitted with a remote control unit to fly the ROV away from the ship and start its dive as the deck crew paid out the tether. Only when it was safely underwater did the pilot on deck pass off the operation to the pilots in the control room.

It was an impressive process, designed to minimize the window of greatest vulnerability for any package passing through the frequently bumpy interface between air and sea. Because of this, they were able to launch and recover in much worse sea states than we could risk with the Johnson-Sea-Link. Also, they weren't limited by battery charge the way we were in the subs, since all the power they needed was transmitted down the tether. It promised the opportunity for extended periods of animal observation. But there was one little hitch: Those observations would have to be made from the ROV control room, which was situated in the bow of the ship—pretty much a guaranteed roller coaster on a 110-foot vessel. There was no choice. I took one last gulp of fresh air and proceeded belowdecks.

The control room was a high-tech, dark space lined with large video monitors in front of four deck-mounted padded chairs like those you'd find in an airplane cockpit. Two of these chairs were for the ROV pilots and two were for the scientists who were supposed to direct the pilots and keep up a running commentary of observations and animal identifications on the audio track of the high-resolution video that was recorded during each dive.

I was in the chair nearest the bow, staring at the monitor directly in front of my face, which was displaying an underwater scene being recorded by one of the cameras on the ROV. Since the motion of the ship was being transmitted down the ROV's tether, the scene was bouncing up and down, but it was completely out of sync with the up-and-down motion I was feeling. It was a formula for intestinal disaster that had claimed many victims over the years[*] and saddled the ship with its unflattering nickname, the *Point Puke*.

As my nausea grew, it was becoming increasingly difficult to focus on the instructions that Robi was giving me. Eventually, when it became clear that mind was going to lose to matter and gastric cataclysm was imminent, I mumbled something about getting a cup of coffee in the mess and bolted for the head. I was fooling no one; Robi and the ROV pilots had seen it all too many times before. After relieving myself of the vile bilge sloshing around in my stomach, I attempted to return to the control room and take up my station as though nothing had happened.

By this time the ROV had reached the bottom, so we started looking around for a good place to set down the bait box. The *Ventana* was facing one of the canyon walls, and our lights revealed a couple of convenient ledges that appeared ideal. I pointed to what looked like a good spot and the pilot started to maneuver the ROV into place. The next hour was an education in the operational limitations of ROVs.

It seemed like every time the vehicle got close to the ledge, the tether would drag it back. From my experiences diving in Wasp, I knew how restrictive a tether could be, but after many years diving in the JSLs I had forgotten. With so much fumbling about, the ROV kept stirring up the fine silt that had built up on the ledge, creating mini dust storms, and we had to wait for them to clear to see what

---

*One of the most famous was Alan Alda, of M*A*S*H fame, who came out on the *Point Lobos* soon after he began filming his wonderful PBS series *Scientific American Frontiers*. He turned visibly green on camera and never filmed another episode of that program at sea.

we were doing. It seemed to take forever, and during that time I had to make a couple more dashes to the head.

Eventually, after many tries, the ROV was lined up properly and the pilot used the manipulator to set the bait box on the bottom, but as soon as the box was released, I watched in stunned amazement as it slid toward the edge of the cliff as if possessed, like one of the kitchen chairs in *Poltergeist*. Another disadvantage of ROVs is that they have no depth perception. It sure didn't look like it, but that ledge had a sloping bottom—a pretty steep one, judging by the speed with which the bait box was exiting the scene.

Fortunately, the pilot managed to snag it with the manipulator before it went sailing off into the abyss, but now we had to find another ledge and begin the whole process again. And after all that, one last fly in the ointment came to light when we finally got the bait box situated: I learned that there was no way to go neutral and shut off the thrusters—something I did all the time in submersibles like the JSLs and Deep Rover. For safety reasons, ROVs are trimmed to be positively buoyant so that if they ever lose power, they will float to the surface. That means that to stay down, their thrusters must be running all the time. So much for being unobtrusive! In the end we got no data.

For my next attempt, my first goal was to beat the seasickness. Unlike the last time, I made sure I was well rested and I took some medication described as the "Coast Guard cocktail," which former *Point Puke* victims swore by and promised wouldn't make me sleepy. As we cruised to our dive site, I stood out on deck staring fixedly at the horizon. I also chose a different location to deploy, a deep site that was farther out to sea, giving me a little more time to get my sea legs. All told, it worked. I felt fine.

This new site also had a smoother bottom, so we didn't have to waste time looking for a flat spot. We just dropped the bait box and backed off a few feet to observe, beginning by comparing what we could see under red light to what was visible under white light. The combination of the red light with the intensified camera worked brilliantly. In fact, because the camera was black-and-white and had automatic gain control,* I had to make sure I took careful notes so I knew which lights, red or white, were on in each recording I made.

With the red lights on, we soon saw hagfish and large sablefish nosing around the bait box. When we turned on the white lights, the sablefish would disperse immediately, while the hagfish remained—not surprising, given that hagfish don't have image-forming eyes and are driven primarily by smell. Given enough time, the sablefish would return when the white lights were on but would remain near the box for much shorter periods. On average, I saw thirty-nine sablefish in each ten-minute viewing period under red light, and only seven under white light. Those were convincing numbers, but it was nonetheless obvious from their behavior that the sablefish could see the red light, because when it first came on after a period with the lights off, the fish would swim away. However, it was also obvious that red light was far less aversive than white light and that using an intensified camera was the key to making it work as an illumination source.

Since it was apparent that there was no way to make an ROV unobtrusive, the next step was to build the battery-powered camera I had envisioned in the first place. Still, funding remained elusive. I ended up kludging the system together with multiple funding sources. The first and most unusual involved something called the Engineering Clinic, an innovative hands-on undergraduate teaching program pioneered by Harvey Mudd College, in Claremont, California. The idea was to give students experience working in teams to solve real-world engineering problems for "clients." To become a client, you needed to submit a proposal, pay a fee of $35,000, and pro-

---

* A closed-loop feedback system in which changes to the input signal, in this case the illumination level, have minimal impact on the output, i.e., the image brightness.

vide all the requested materials. I didn't have nearly that much money in my internal budget, so I had to convince Harbor Branch to pitch in the fee, and I used the sale of some of my images of deep-sea animals to cover the equipment purchases.*

As far as I'm concerned, hands-on problem solving is the ideal way to learn, and this project was a splendid example of how motivating it can be. The Harvey Mudd students were receiving a practical, multidisciplinary education by meeting a real-world engineering challenge while at the same time hearing about deep-sea biology. Being involved in something so exploratory was exciting for them and helped sustain their enthusiasm through long hours. There was a lot to do, with many challenges to overcome, but they got it done, creating a computer-controlled camera/recorder/illumination system that worked on the bench top.

With the bare bones of the system built and the red-light data from MBARI as proof of concept, I went to NOAA for the $15,000 I needed to put the camera/recorder into an underwater housing and design and build a frame to deploy it on the bottom. Once again, MBARI helped out by providing the underwater batteries I needed to run the system, and, most important, they provided the ship time necessary to test it in the field for the first time, in 2002. This was the mission that Dave Clark had asked to film, the first field tests of the Eye-in-the-Sea.

As an example of cutting-edge engineering for ocean exploration, it was an ungainly-looking contraption. The tripod, built out of aluminum tube stock, was over seven feet tall, with the battery on the bottom and the camera/recorder in a cylindrical underwater housing mounted just above it. The red light was mounted as far off-axis from the camera as feasible, at the top of the tripod, in order to reduce the backscatter that would degrade the image with light reflecting off particles in the water.

---

*This is a scientist's version of setting up a lemonade stand.

The day of the launch was sunny and clear, the waters so calm no one had to worry about getting their sea legs. Once we reached our deployment site, the ROV crew made short work of lifting the EITS off the deck and carrying it two thousand feet deep, where they deployed it and the bait box with such speed that we had plenty of time to go exploring afterwards. Clark wanted to film as much as possible while we were out there. Robi had come along as co–chief scientist and was happy to give us a tour of his backyard while showing off the capabilities of the ROV to capture high-resolution imagery of fragile gelatinous organisms. The whole day went perfectly.

The next day, not so much. It started when my borrowed pinger on the EITS failed to provide a signal we could home in on. It's an awfully big ocean, and I felt my anxiety ramping up as we searched unsuccessfully. Eventually the ROV picked up a target on sonar that led us to it, and I experienced a rush of relief when I saw it standing just as we had left it. The bait box, on the other hand, was now surrounded by clusters of crabs and sea urchins and there was a writhing mass of hagfish slithering in and out of the plastic mesh holding the bait. It looked as though there must have been plenty of on-camera action to record.

It was when we got the EITS back on deck that things went seriously south. As Clark filmed my every move, I assiduously tried to ignore him and his camera as I hooked up a long cable between the EITS on deck and my laptop, back in the lab. I said a little silent prayer to Saint Murphy—*Please have mercy on me!*—and typed in the command to retrieve whatever imagery the camera had collected. *Nothing. Nada.* The system refused to talk to me, no matter how nicely I asked. After several failed attempts, I went back to the deck to check the camera connection and that was when I saw it: water sloshing inside the camera dome.

Only once before had I seen that sickening sight. It was early in my career, shortly after I had started at Harbor Branch. I had been in need of a supersensitive light meter that I could use on the JSL.

Nothing of the kind existed, and I had to convince two funding agencies (ONR and NSF) to share the cost of its development. On an early test dive, it failed, and one of the sub crew, Jim Sullivan, a.k.a. Sulli, spotted the water in the optical dome and delivered the devastating news.

At that time, I had felt like my career was hanging by a thread as I stared dumbfounded at the water sloshing back and forth. Sulli, who was a font of both philosophy and humor and whom I thought of as my personal Yoda, stood quietly next to me for a couple of minutes and then drawled, "You know, success in life depends on how well you handle plan B. Anyone can handle plan A." It took a while for that to sink in, but once it did, it turned out to be exactly what I needed to hear. I wrote those words down and posted them on my office wall, which is why I knew just what to say when Dave Clark shoved his camera in my face to get my reaction to the EITS's flooding.

Robi, too, knew just what to say. Once we were back on the dock, he made an eloquent on-camera speech about how we have to expect to fail every now and then and how David Packard said, "If you're not failing occasionally, I'll think you're not reaching far enough. I want you to reach as far as your imagination will allow you."

Generally speaking, when people start making speeches about your courage to fail—on national television, no less—it's not a good sign. The fact that this was exactly the worst of the worst-case scenarios I had envisioned did not help at all. I felt like I had been gut-punched, and I couldn't even claim it was a sucker punch, because I'd volunteered knowing this could happen! It wasn't just the mortification of an extremely public failure; there was also the far more concerning worry that this could endanger any hope of future funding for the EITS. I knew I needed to make it right, and fast.

I had just three months to turn things around before my next mission. First, I asked Clark if he would film this (hopefully) comeback expedition. He said he was running out of time and money, but if I managed to get some good footage from the EITS he would try to include it.

This was going to take some serious scrambling and scrounging. Fortunately, I had an ally. Lee Frey was a young ocean engineer who had started at HBOI in 1997 as an eager intern and worked his way up to senior engineer in less than five years. Besides having an obvious passion for deep-sea exploration, Lee had an incredibly valuable talent for adapting the engineering to the budget at hand, which in this case was rapidly approaching zero. Somehow, whenever a problem arose, he kept finding workarounds that didn't break my pathetic bank. But still, it all felt pretty dicey, so when the launch of the next mission rolled around, I wasn't terribly confident about the outcome.

Clark didn't come along on the ship this time. I was initially relieved he wasn't going to be there, because if it all failed again, I certainly didn't want to share that moment. However, as fortune—and Murphy—would have it, this time everything worked.* The whole operation played out exactly as it had before, only this time, when I plugged in the camera cable and typed in the command to retrieve images, after a brief but seemingly interminable pause, video sequences started appearing on my laptop. It's hard to imagine a better feeling in the world than that moment of pure victory when you manage to pry open a door that has remained stubbornly shut for so long. I was seeing the deep sea in an entirely new way—without scaring the animals I wanted to observe!

The recordings were mesmerizing. I felt like a kid again, watching the action around home base while remaining hidden. There was no bioluminescence, but there was plenty of action, with fish and sharks swarming around the bait. Even better was seeing that imagery on national television when Clark's documentary *Science of*

---

*This is actually one of the innumerable corollaries to Murphy's law: When things go right, nobody notices (or in this case is there to see).

the Deep: Mid-Water Mysteries aired in early 2004 on the Science Channel. The show won Clark and his coproducer, Sue Norton, the National Academies Communication Award,* "for showing the importance of engineering in scientific exploration." They did show the initial failure of the Eye-in-the-Sea, but they also showed its ultimate success.

Being willing to fail is one of the job requirements if you want to explore any kind of frontier. There are many quotes on the subject. My favorite is Winston Churchill's: "Success is stumbling from failure to failure with no loss of enthusiasm." Your passion needs deep roots if it's going to survive the "slings and arrows of outrageous fortune." For some, their obsession stems from wanting to be the first to reach a place, like the moon or the bottom of the Mariana Trench. It's a powerful drive that has resulted in valuable technology development.

But for others, their fervor involves unmasking the hidden secrets of the natural world, an aspiration that doesn't necessarily require traveling to the farthest reaches of the globe. The invention of the microscope revealed a previously hidden world, including the first microorganisms ever seen. How utterly amazing to discover another world within our world.

What world within our world might be revealed to us in the vast reaches of the dark ocean if we can simply learn to explore it without scaring life away?

---

*A project of the National Academy of Science, the National Academy of Engineering, and the Institute of Medicine, funded by the W. M. Keck Foundation, to help the public understand topics in science, engineering, and medicine.

Chapter 11

THE LANGUAGE OF LIGHT

Bumbling around a frontier in the dark, both literally and figuratively, one learns to expect calamity as the natural order of things. Yet a combination of insatiable curiosity and optimism sustains forward momentum in the face of what seem like endless setbacks. Maintaining optimism takes work, often requiring a concentrated focus on small successes. As a result, when a major success materializes, it can be kind of overwhelming, even if it was the very thing you were working toward. In my wildest imaginings I never envisioned anything as amazing as what occurred the first time I took the Eye-in-the-Sea on a major expedition.

MY THRILL OF victory after that first successful deployment of the EITS in the Monterey Canyon was short-lived. During my initial experiments, I hadn't been terribly surprised that the sablefish we observed from the ROV seemed to be able to see the red light emitted through the red plastic filters snapped over the ROV's white lights; I had measured the transmission through those filters and knew that some of the shorter wavelengths—blue light—were sneaking through. But for the EITS I had switched to red LEDs, which are often functionally described as emitting monochromatic light. Most deep-sea fish are also described as monochromats, meaning they *see* only one color. Since that one color is usually blue (between 470 and

495 nm*), I was hopeful they wouldn't be able to see the light emitted by the red (660 nm) LEDs. But with careful analysis, it became clear that they could.

Although some of the fish didn't seem to respond when the red lights came on, others would gradually adjust their swimming patterns, veering away from the lights. In a few cases, it wasn't even all that subtle: When the lights came on, they bolted. For my next try, I pushed the illumination even further, using far-red LEDs (680 nm). The penalty for this was that the longer the wavelength, the more poorly it transmitted through seawater. The resulting video recordings were much dimmer and harder to analyze, but the outcome was the same: The fish were *still* seeing the light.

Calling LEDs monochromatic and fish monochromats is misleading. Even though most deep-sea fish have only one visual pigment, that pigment can absorb a whole range of colors—the fish just can't distinguish one from another. Their vision is like a black-and-white camera through which varying colors are simply seen as different shades of gray. For the sablefish, maximum sensitivity is to blue light (491 nm), so the color blue would be perceived as white, green of equal intensity would be seen as light gray, yellow would be medium gray, orange would be dark gray, and red would be seen as close to black.

Plotted on a graph, this visual sensitivity looks like a very broad bell-shaped curve, with its peak sensitivity in the blue but with the base of the bell, where sensitivity gradually drops toward zero, extending to the left into the short wavelengths down past 400 nm to the ultraviolet, and surprisingly far to the right into the long wavelengths all the way past 600 nm to the oranges and approaching the reds. The emission of the red LEDs, on the other hand, looks like a

---

*Wavelengths of visible light are measured in nanometers (nm). One nanometer is one-billionth of a meter. Visible light ranges from 400 nm (the blue end of the spectrum) to 700 nm (the red end).

very narrow bell-shaped curve with its peak at 680 nm. If you plot these two curves on the same graph, they don't appear to overlap, but if you greatly expand the vertical axis* and zoom in on that stretch between the two peaks, you can see that in fact they do.

It seemed like I was between a rock and a hard place. The width of the visual pigment absorption spectrum, on the one hand, and, on the other, the extreme attenuation by seawater at wavelengths outside that range, at the red end, were making it extraordinarily difficult to see without being seen.

IMMEDIATELY AFTER THE successful trials of the Eye-in-the-Sea in 2003, I wrote a proposal to NOAA's Office of Ocean Exploration and Research for an expedition to explore the ocean in a whole new way—one that took into account the visual capabilities of its inhabitants. What might we find that had never been discovered before, if we could see without being seen?

My goal was to have the EITS play a major role in this expedition, which was scheduled to take place in the Gulf of Mexico in 2004. The trouble was that, as the date of the expedition approached, I was still struggling with making the illumination unobtrusive.

My inspiration for a solution came from the extraordinary stoplight fish, whose bioluminescence I had measured using the Optical Multichannel Analyzer back when I first started investigating the colors produced by various light producers. Like many deep-sea fish, it had a flashlight next to its eye that could emit blue light. But it also had a much bigger flashlight under its eye that shone red. The remarkable thing about the stoplight fish is not just that it can emit red light, but that it can see it! This means that the fish has sniper-scope vision that potentially allows it to see without being seen.

What an incredible advantage: to be able to home in on prey that

---

* Or, for those who are more mathematically inclined, if you plot them on a logarithmic scale.

can't see you and communicate with potential mates without revealing your presence to your predators! One of the striking features of this red-light organ is that it has a very sharp cutoff filter* over it. This filter shifts the color of the raw light produced by the light organ from a bright reddish orange to a much dimmer infrared. It is an impressive color shift, and at the time I made those measurements, I had been struck by how much light energy the fish sacrificed to blot out those shorter wavelengths. The selection pressure to be sneaky is clearly enormous.

Taking my cue from Mother Nature, I decided to emulate the stoplight fish by using cutoff filters in combination with the red LEDs, which would hopefully make the illumination less visible to the fish. I was excited to learn that some new, higher-power 680-nanometer LEDs had just come on the market, which would allow me to more than compensate for the energy I was sacrificing by using filters. I had hoped to test this new illumination system on another MBARI deployment. Unfortunately, the cutoff filters were a special-order item with a long lead time, leaving me no window for further testing before the NOAA expedition.

Besides planning to use the new filter system, I also had a new trick up my sleeve. I was hoping to take another stab at talking to the animals. Two decades before this, while diving Wasp and Deep Rover, I had tried to elicit some kind of bioluminescent communication by flashing a simple blue light that I had attached to a pole. I was now convinced that the reason this light wand had failed to elicit any reactions was that I wasn't being stealthy enough. If the new illumination system worked as I hoped, I would finally have a way to test this theory. Also, I had been on dozens of missions and learned a lot about the specificity of bioluminescent displays since those

---

* Like a low-pass sound filter that passes sound frequencies lower than a selected cutoff while blocking higher frequencies, this filter passed low-frequency red light and blocked higher frequencies, including blue, green, yellow, and orange light.

early days. So this time, my attempt to talk to the animals wasn't going to just feature a single blue light but rather would imitate particular displays; in other words, I'd be using a language they might recognize.

The most spectacular display programmed into the device mimicked that of the deep-sea jellyfish *Atolla wyvillei*. Both in the light and the dark, this is a magnificent creature. With the lights on, it looks like a scarlet sunflower with long crimson tentacles protruding out from between its translucent red petals, called lappets. With the lights extinguished, its bioluminescence can take an intriguing range of responses, depending on the strength and location of the stimulus applied. Touching a lappet may induce it to squirt out a thin streamer of light that hangs in the blackness to distract a predator—"Hey, look over here!"—as the *Atolla* covertly escapes into the darkness. A gentle bump on the bell causes a brief, dim pulse of light local to the contact, which I interpreted as a visual "Ahem, this space is occupied." A much more prolonged stimulus, as might occur if the jellyfish was being munched on by a predator, causes it to pull out all the stops and produce a pinwheel display that swirls round and round its surface in waves of brilliant sapphire blue.

The fact that the pinwheel is such an eye-catching, prolonged spectacle makes it an ideal candidate for a burglar alarm—a scream for help, using light instead of sound, to draw attention to an attacker. The superbly sensitive eyes of deep-sea dwellers are tuned to detect any telltale flash that could lead predator to prey. The prey, in this case, is not the jellyfish but its attacker. The jellyfish "scream" is a plea for rescue—"Help! Come eat this guy before he eats me." I thought a display like that could potentially attract predators from hundreds of feet away. If it did, it would have the double benefit of helping substantiate its function as a burglar alarm while drawing large predators into the field of view of the EITS.

This new lure, dubbed the electronic jellyfish, or e-jelly for short, was fashioned as a ring of blue LEDs on a circuit board that was then cast in clear epoxy to make it waterproof. For the mold, we used a round plastic food container, which had a medusa-like shape and appearance—so long as you ignored the impression of the letters across the top of the bell spelling out ZIPLOC.

My goal was to place the EITS at some deep-sea oasis—a biologically rich patch of ocean floor that large predators might be likely to patrol—and leave it there for periods of a day or more to see what might come around. The oasis I selected was one that I had wanted to visit since I first learned about it. Known as the Brine Pool,* it's an underwater lake in the northern Gulf of Mexico. The idea of an underwater lake is an Alice in Wonderland concept that is difficult to get your head around. It's definitely one of those things you have to see to believe.

Back when dinosaurs roamed the Earth, the Gulf of Mexico looked very different than it does today. It was smaller, with a narrower opening to the ocean. Periodically it would dry out, creating thick layers of salt. Plate tectonics eventually caused the gulf to open up and flood. Sediments then built up on the seafloor. In places where those ancient salt deposits poke through the sediment, the salt dissolves, forming super-salty seawater called brine. Brine is heavier than regular seawater and therefore collects in pools, with distinct shorelines, on the floor of the gulf.

As if that's not weird enough, in some cases, hydrocarbon deposits co-occur with the brine, resulting in chemosynthetic communities similar to those found around hydrothermal vents, only there is no extreme heat involved, so these are called cold seeps. Life flourishes here in the absence of sunlight because energy-rich com-

---

*Although there are multiple brine pools in the Gulf of Mexico, this one is the most famous and most studied and is therefore capitalized.

pounds like methane, hydrogen sulfide, and ammonium seep up through the seafloor, providing sustenance for organisms like tube worms* and giant mussels that manage to subsist on these compounds with the help of symbiotic relationships with chemosynthetic bacteria. One such community around the Brine Pool has the largest accumulation of chemosynthetic mussels found so far in the Gulf of Mexico.

That I was actually going to get to see this fantastic place with my own eyes was thrilling just by itself, but I was also entertaining high hopes for what new discoveries might arise out of the Eye-in-the-Sea rigged with its simulated stoplight-fish illumination system and e-jelly lure. Plus, my collaborators on this adventure, a mix of international experts in optical oceanography and visual ecology, shared my passion for understanding what part light plays in the lives of animals in the ocean. As a bonus, they were a lot of fun to be with at sea.

All told, there were sixteen of us in the science party. Key players included Tammy Frank, who was serving as co–chief scientist with me and studying the visual sensitivities of animals that we captured in the dark; Sönke Johnsen, who had moved on from being my postdoc to a tenure track position at Duke University and who planned to analyze the camouflage strategies of deep-sea bottom dwellers; Justin Marshall, an expat Brit who had moved to Australia and would work with Sönke to study what role polarized light might play in deep-sea visual ecology; and Erika Montague, my graduate student at the time, who was working with me on the Eye-in-the-Sea experiments. Sönke, Erika, and Justin shared a similar slightly twisted sense of humor, and based on past experience, their practical jokes

---

*The ones in the Gulf of Mexico have smaller-diameter tubes and smaller flowers than the ones discovered in the Pacific, but they can form into very large bundles called bushes.

on one another were likely to provide considerable entertainment for the rest of us.*

The mission plan was to visit four dive sites during the ten-day expedition. The Brine Pool, farthest from our starting point in Panama City, Florida, was last on the schedule. To undertake all the planned experiments and give all six principal investigators opportunities to dive in the submersible, we were scheduled to make three dives a day with the Johnson-Sea-Link rather than the usual two. It was a densely packed, complicated, and optimistic schedule.

A mere eight hours into the mission, the weather turned nasty. The wind was blowing at twenty knots, and predictions were that it was only going to get worse. Rather than waste precious dive time sitting around waiting for it to clear, we decided to reverse the schedule and make the long transit to the Brine Pool, which was about 150 miles southwest of the Mississippi delta, and hope that the weather would clear up (as the reports were predicting) by the time we got there.

We arrived onsite late afternoon the next day and, miraculously, the weather was glorious. The first dive hit the water just after four P.M., with the pilot and me up in the bubble while Erika rode in the rear dive chamber with a subby.† To stay on schedule, we needed to get in two dives before midnight. It would be tight. We had half an hour to reach bottom, just below 2,100 feet, an hour to deploy the EITS along with some animal traps, then another half an hour to return. As we descended, I watched eagerly for signs of bioluminescence. The first flashes appeared at around 1,000 feet, at which

---

*At the end of one expedition, Sönke and Erika stole the blades out of Justin's electric razor right before he left the ship. Trying to shave on the long flight to Australia, he was puzzled when it wasn't working. A couple of days later, he received a ransom note with the blades displayed against that day's newspaper and photoshopped wearing sunglasses and smoking cigarettes.

† *Subby*—slang for sub-crew member.

depth I could still distinguish dim blue light overhead. By 1,200 feet, that blue ceiling had turned charcoal gray before fading to black at 1,800 feet, when the pilot flipped on the sub's lights.

We continued our descent down past 2,100 feet, where the edge of the Brine Pool loomed into view. The pool had a distinct surface, a consequence of the refractive index difference between its dense salty brine and the much less salty seawater overlaying it. It was a disorienting sight—a lake within a sea, with a conspicuous shoreline covered in giant mussels. These were at least twice as big as any I'd ever seen, and they were packed in so tight, there were a couple hundred per square foot. Their colors were a kaleidoscopic mélange of browns—walnut, rust, peanut—and charcoal grays, with flecks of white from open shells, while the lake was nearly black, grading to a dark mossy aquamarine in the direct splash of the sub's lights.

As we approached, I saw a hagfish swim down in front of the sub, pass through the surface, and disappear into the lake, only to reappear and swim away. Other, less primitive fish that had tried this apparently hadn't fared so well in the salty brew, as evidenced by a couple of carcasses floating on top of the lake.* I asked the pilot what would happen if we tried to sink into the brine. "We can't get through—it's too dense," he answered and then proceeded to demonstrate by setting the sub down on top of the pool. Bizarrely, his maneuvers on the lake's surface created slow-motion waves that lapped against the shore. The whole scene had a fantastic, alien spookiness about it.

We cruised over the surface of the pool and then up along the western edge, looking for a good spot to deploy the EITS. Erika and I agreed that the mussel beach† that formed a wide ring around the pool looked too lumpy to set the camera on safely, so we ended up

---

* Obvious proof that too much salt in your diet is not healthy.

† Little-known fact: Muscle Beach, in Venice, California, was first known as Mussel Beach, back when it was famous for bivalves instead of biceps.

MBARI's remotely operated vehicle deploying the Eye-in-the-Sea. E. WIDDER

The deep-sea jellyfish *Atolla wyvillei*. E. WIDDER

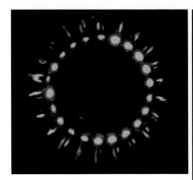

The bioluminescent burglar-alarm display of *Atolla wyvillei* is a pinwheel of light that swirls around the edges of the bell (above). At right is the e-jelly used to imitate that display.
E. WIDDER

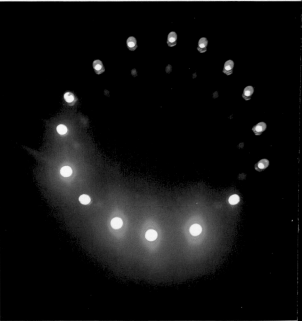

Eye-in-the-Sea on the shore of the Brine Pool. E. WIDDER

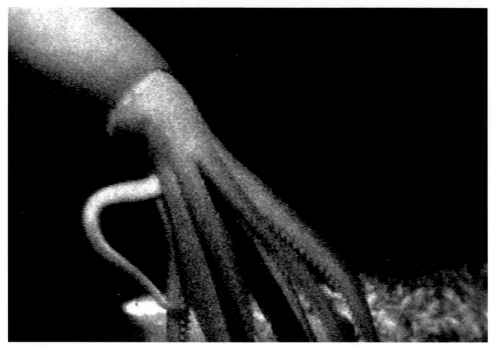

Just eighty-six seconds after the e-jelly was activated, the Eye-in-the-Sea recorded this squid, a creature so new to science it could not be assigned to any known scientific family. E. WIDDER

The bamboo coral *Keratoisis flexibilis* releases astonishing amounts of slime and lights up like a Christmas tree when you rub against it (inset). E. WIDDER

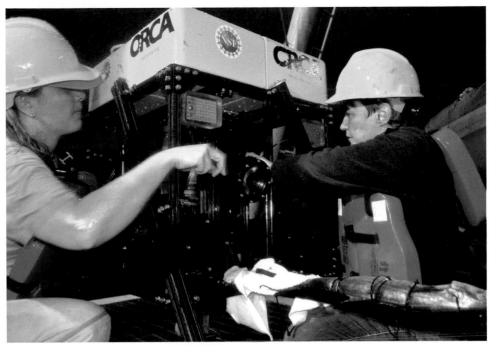

Brandy Nelson and Edie Widder readying the Medusa for deployment near the site of the BP oil spill. SYLVIA EARLE

The three scientists on the giant squid expedition (from left): Tsunemi Kubodera, Steve O'Shea, and Edie Widder. LESLIE SCHWERIN

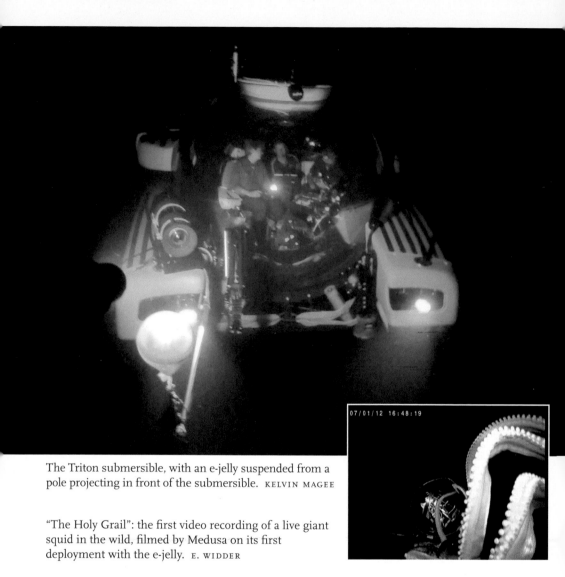

The Triton submersible, with an e-jelly suspended from a pole projecting in front of the submersible. KELVIN MAGEE

"The Holy Grail": the first video recording of a live giant squid in the wild, filmed by Medusa on its first deployment with the e-jelly. E. WIDDER

Attack of the Kraken: The giant squid homed in on the e-jelly and then attacked the very big thing next to it, the Medusa. E. WIDDER

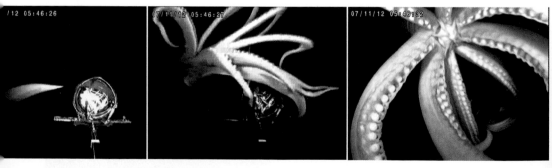

Close-up of the giant squid, showing its striking coloration, like brushed aluminum, and very large eye. COURTESY OF NHK/ NEP/DISCOVERY CHANNEL.

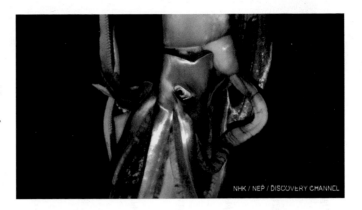

NHK / NEP / DISCOVERY CHANNEL

Enypniastes eximia, first described by ROV pilots as headless chicken fish, are swimming sea cucumbers. Their bodies are covered with sticky blue bioluminescent particles that rub off, turning any predator into a glowing bull's-eye. ABOVE, COURTESY OF NOAA PHOTO GALLERY. RIGHT, E. WIDDER.

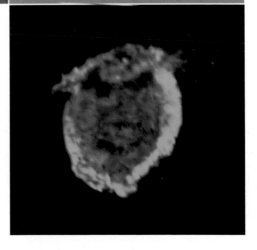

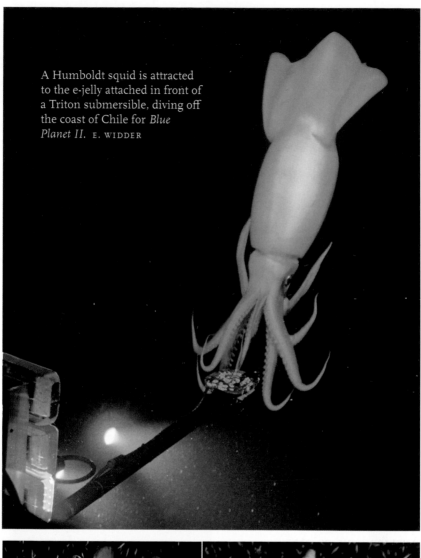

A Humboldt squid is attracted to the e-jelly attached in front of a Triton submersible, diving off the coast of Chile for *Blue Planet II*. E. WIDDER

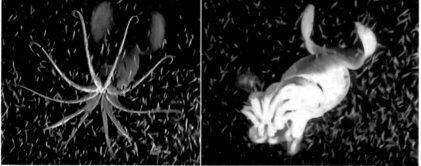

Two still frames from Deep Rover pilot Toby Mitchell's iPhone video of the Humboldt squid basket-feeding on krill. TOBY MITCHELL

picking a spot just outside the mussel bed at the northeast end of the pool. This was our first time deploying the EITS from the submersible, so I was worried it might refuse to slide off the rig that was holding it on the front of the sub. But thanks to the robotic arm and the pilot's skill, the deployment was flawless.

Until we recovered the camera on a subsequent dive, it was impossible to know what the video was picking up, so I had to guess at the field of view, but we tried to line it up with the mussels in the foreground and the edge of the brine pool in the background. We placed the e-jelly, which was on a cable attached to the camera bottle, on top of the mussels, a few feet in front of the camera next to a net bag full of bait. The goal was to draw as many critters as possible into the camera's view. We also put the animal traps in the same vicinity, with the hope of seeing how the animals reacted and interacted with them. Then we cruised back across the pool to its southeast corner, where there were large bushes of tube worms. We placed the other two traps there and then, incredibly, it was already time to surface.

Both Erika and I desperately wanted to stay and explore this amazing place, but we mollified ourselves by asking the pilot to turn out the lights during the ascent so we could revel in the spectacular display of bioluminescence. It was breathtaking. We could stimulate it either mechanically, by bumping into it, or photically, by exposing it to a brief flash of light from our flashlights or camera strobes. Anywhere we pointed our flashlights, we would see fragile filigrees of fluff light up and then fade away. And if we used the camera strobe, the effect was even more remarkable, with whole galaxies of luminescence flashing back at us in unison from all around the sub.*

We left the camera down overnight. First thing the following morning, I went to retrieve it. Tammy rode up front and I was in the back, operating a spectrometer that we were using to measure the

---

*The cause of this flashback phenomenon is unknown. More about that later.

penetration of downwelling sunlight into the depths. It was a bulky piece of gear for the small confines of the dive chamber, leaving little room for me and the subby. As we descended, I made a series of measurements that beautifully charted the narrowing spectrum of light characteristic of the deep ocean.

Having stored the last of the spectra on the computer, I was trying to rearrange the gear to give myself enough room to stretch my legs when Tammy yelled into the headset, "Look at that!" I checked the video monitor that showed the camera feed from the front of the sub. Filling the screen was a giant sixgill shark, and it was swimming right past the Eye-in-the-Sea. The word *giant* felt like an understatement. The pilot shone the scaling lasers* on it and estimated it was almost *fifteen* feet long. Since the EITS was still in record mode, I had every reason to believe we might have video of this goliath, hopefully before the submersible arrived on the scene. I couldn't wait to get it back on deck and see what we had recorded.

We had nothing. Not a single frame. It was a crushing disappointment. We consoled ourselves that at least it hadn't flooded. Erika and I immediately went over the system piece by piece but could find nothing mechanically wrong and eventually concluded that one of us (who shall remain nameless) had inadvertently loaded an old configuration file. Sönke had once said that we all lose at least ten points of IQ at sea,† and this event supported his theory. We immediately instituted a policy of both of us inspecting the setup for each deployment and, after triple-checking that we had the right configuration file loaded this time, we had the camera ready to go down again right after lunch. We had programmed it to collect video

---

*Two red lasers mounted in parallel a known distance apart—for example, 10 centimeters—so when you see two red dots on the shark's flank you can use that known distance to calculate the full length of the shark.

† Possibly because on a rolling ship a portion of your brain is perpetually diverted to calculating which way is up?

through to the next morning, when we would make our last dive at the Brine Pool.

This final dive had Justin in the front and one of the grad students in the back. They reported no sign of the sixgill shark, which was disappointing, but they recovered the camera without incident and were back on deck with the EITS before ten A.M. Erika and I immediately checked the camera and were relieved to see that this time we had video, but there was no time to look at it because we had to quickly download the memory and prep the camera for the next deployment, at a place called Green Canyon.

This time, Sönke sat in the front while I worked the spectrometer in back. It was one of the few times in my life that I was anxious for a dive to be over. After so many years of trying to peer into the darkness, I had reason to believe we might finally have succeeded, and I wanted to get back and find out what was there. We deployed the camera near some tube worms and then hovered a hundred feet above it to try to see the e-jelly come on. We had programmed it to start displaying immediately this time and, sure enough, it came on and was readily visible from that distance. I had real cause to hope that it might draw in visual predators from a long way off.

As soon as we were back on the ship, I adjourned to the lab to review the Brine Pool video. The good news was that everything seemed to be in focus. The bad news was that neither the e-jelly nor the bait bag was in the field of view. It looked as if the tripod might have settled back into the mud, tilting the camera up so that both the bag and the e-jelly were just below the bottom of the frame.

Nonetheless, I could still see some of the mussels and the edge of the brine pool. Even better, I could see fish and giant isopods swimming around, seemingly oblivious to the camera lights. To almost anyone else, I'm sure it looked like a pretty uninteresting scene, but I was on the edge of my chair, my eyes riveted to the monitor. I was finally gazing into this other world, and I was invisible to its inhabitants. That meant that at any moment I might see something no one had ever seen before. For me, this was akin to discovering the entrance to King Tut's tomb, and although I didn't know it at that moment, I was about to uncover the golden sarcophagus.

For this first deployment of the EITS, I had delayed activating the e-jelly until four hours had passed so I could have a good, long look at undisturbed behavior. I had gotten to the section of the video where the e-jelly was activated when, just *eighty-six seconds* after it came on, the video screen filled with what looked like an enormous squid. I bolted out of my seat and whooped so loud that people came running from all over the ship to see what I was shouting about. We all watched it over and over again.

It was a strange-looking squid. The oddest thing about it was its tentacles, which were short and muscular instead of long, thin, and stretchy. The squid looked to be attacking the e-jelly just below the bottom of the frame. As it withdrew from the failed attack, apparently by flapping its fins, which we could see only partially at the top edge of the frame, it flexed one of its tentacles out to the side. The appendage was so thick and short, it looked like one of the arms, except that it was lighter-colored, lacked suckers, and was two-thirds the length.

One of the great blessings and curses of modern oceanography is that we now have email at sea. That access meant that, instead of having to wait to return to shore, I could send the video clip to squid experts at the Smithsonian, who got back to me almost immediately with the amazing pronouncement that this squid was *completely new to science.* It wasn't just a new species or a new genus, but possibly an entirely new family!* The full significance of that took a while to sink in, but I was sure of one thing: I could not have asked for a better proof of concept. The discovery of a new animal alone would have

---

*Recently, it was tentatively given the name *Promachoteuthidae* based on three captured juveniles with similar characteristics.

been enough to declare this expedition a massive success, but we weren't done yet.

As we all sat around reviewing the video of the mystery squid, we were bemoaning the fact that the e-jelly and bait bag weren't in the field of view. I wanted to be able to definitively say that the squid had been attracted by the e-jelly, and I also needed a scale reference to properly size the squid. I guessed it was about six feet long, but that was based on how far away I estimated the e-jelly was from the camera. The only solution in the short term seemed to be to make every effort to place the e-jelly squarely in the camera's field of view for each of the remaining deployments. However, on the next deployment we faced the same problem: The e-jelly and bait were out of sight, just below the bottom of the frame.

We all agreed that the solution was to attach some kind of extension to the camera frame that would hold the e-jelly at a fixed distance from the camera. Because of the very tight clearance during launch and recovery between the EITS, loaded on the front of the JSL, and the aft end of the ship, it would need to be something that we could fold down after the EITS was on the bottom and then fold up for recovery. I had no doubt the engineers back at Harbor Branch would find a way, but that would have to wait until the next expedition.

At least that was my thinking as I drifted off to sleep that night, but the following morning a new option was presented to me by Justin and Erika, who had stayed up all night fashioning a Rube Goldberg solution. They had found an aluminum ladder that they convinced the captain to let them cut up. It was now attached to the bottom crossbar on the EITS frame with ring clamps so it could be rotated from horizontal to vertical, where it locked into place with a spring-mounted hook. The e-jelly itself was attached to a separate short section of the ladder, canted in such a way that it was positioned in the bottom center of the frame. It looked sketchy, but it was actually a pretty solid piece of engineering, as was proven on its first deployment, when it worked exactly as advertised. The big moment came after the EITS was placed on the bottom and the sub's robotic arm reached out to release the spring holding the ladder up. The frame rotated down so majestically, when I saw the video of the deployment, I wanted to score it with Strauss's "Sunrise" fanfare, used in *2001: A Space Odyssey.**

We put it down for only a few hours on that first deployment, just to make sure it worked. When we looked at the video, we decided the ladder was a little too high in the field of view and too reflective, so we adjusted its position and spray-painted the whole thing black. The next time we deployed the EITS, we not only attached a bait bag next to the e-jelly but we tie-wrapped fish heads to the ladder rungs. It was a deep-sea sushi platter that we hoped would lure in some exciting guests. Two days later, when we got it back on deck, it was evident that we had indeed attracted visitors to the buffet, because the bait bag was missing and there were scratch marks on the paint all around the fish heads.

This time there was a crowd around the video monitor as I reviewed the recording, and I wasn't the only one whooping. There were hake fish and giant isopods† munching on the bait. A rosefish, which we dubbed the "peekaboo fish," kept coming in to look at the e-jelly every time it came on, and, for the grand finale, another giant sixgill shark loomed up out of the darkness, used its head to shove a giant isopod out of the way, and then opened wide to start chowing down on the bait. We posted these videos on the web page that NOAA had set up to highlight our expedition and later learned that they had created a lot of excitement.

For the proposal to NOAA, I had come up with the expedition name Deep Scope to emphasize that we were going to be attempting

---

* Just the opening bars. Not the full hour of *Also Sprach Zarathustra*. It was majestic, not glacial.

† Picture a pill bug the size of a toaster oven.

a different way of exploring, using new technological eyes to peer into the depths and to provide a novel perspective that focused on what the eyes of animals were adapted to see.* Deep Scope 2004 was so successful that NOAA funded our team to pursue this line of inquiry for subsequent missions in 2005, 2007, and 2009, and the Eye-in-the-Sea played a significant role in each of these expeditions. It seemed as if every EITS mission produced new revelations about the nature of light and life on the deep seafloor, but one of the most thrilling for me was the 2007 expedition to the Bahamas where I finally managed to talk to the animals.

We did replace Justin and Erika's clever contraption with a more streamlined system for holding the e-jelly and the bait, but as homage to their ingenuity, we continued to refer to it by the acronym that Sönke had suggested: CLAM, for "cannibalized ladder alignment mechanism."

The e-jelly was designed to imitate a number of different displays. One of these used a single blue LED to produce a rapid repetitive flash. On the 2007 expedition, there were several instances where this triggered an elaborate response that appeared as a sequence of dabs of liquid light that were released in speedy succession by something—I'm guessing a shrimp—swimming fast in a spiral. The effect was like adding an extra corkscrew flourish to the emission with an emphasis that had a *Take that!* feel to it. The e-jelly would fire off a series of short flashes and something—often several somethings—would respond. It was a conversation with light. I had no idea what I was saying, but I think it was something sexy, because the response bore a striking resemblance to the sexual displays of sea fireflies.

In some cases, the strings of light that sea fireflies (bioluminescent ostracods) produce rise up off the bottom. In other cases, they may be horizontal or diagonal, like drifting strands of glowing pearls.

---

* Not developing a new form of proctology, as alleged by Justin, Sönke, and Erika.

The spacing, intensity, placement, and size of the pearls represent a complex code that permits females to identify potential mates of their own species. It is thought that all this marvelous complexity is an evolutionary response to crowding, because it enables several different species to coexist without confounding the mating game.

Obviously, using bioluminescence to attract mates has a potential downside, since the light might easily also attract predators. Therefore, the evolution of displays that involve the release of long-lasting clouds of light, where the emitters can remain physically separate from their light emissions, is enormously advantageous. With sea fireflies, the females of a particular species have co-evolved to intersect the trajectory of the male's swimming pattern, which she calculates based on the bioluminescent landing lights he so obligingly generates.

You don't need a submersible to experience this particular light show. In fact, you don't even need scuba. Because it occurs in water shallow enough to observe while snorkeling, you just need to hop in the ocean shortly after sunset and wait for the performance to begin. This used to be a common occurrence around the Florida Keys, where, sadly, it has largely disappeared as a result of pollution. Nevertheless, there are still plenty of places throughout the Caribbean where it happens year-round, so you should definitely take the opportunity to see it if at all possible. The experience of being surrounded by hundreds of these displays is like being immersed in a symphony of light. It's definitely worthy of your bucket list. Revel in it while you still can.

PART III

HERE BE DRAGONS

Nothing in life is to be feared, it is only to be understood.
Now is the time to understand more, so that we may fear less.

—MARIE CURIE

Chapter 12

THE EDGE OF THE MAP

We are all explorers. Each one of us is born into this world as a stranger in a strange land. Our explorations shape our understanding of the world around us. As a baby, when you crawled away from the safety of your mother's arms to see what was beyond the next corner, you were satisfying a natural instinct to find something new. Exploring the edge of the map, knowing that at any moment you might unveil one of nature's hidden secrets, is such a deeply visceral thrill that it feels primal.

The happiest people I know (a category in which I include myself) are those who have managed to hang on to a childish sense of wonder at discovering new things. But hanging on is not always easy. Too often, the world is served up as a collection of facts to be learned, rather than grand mysteries to be solved. The ocean holds and hides so much dizzying complexity and wondrous weirdness that there is no end of puzzles to entice explorers. Some of the best of these involve how light—both sunlight and living light—has shaped life in the ocean.

Early investigators examining net-collected specimens at the beginning of the last century were so utterly baffled by what they found there that they declared, "Nothing seems more hopeless in biological oceanography than the attempt to explain the connection between the development of the eyes and the intensity of light at different depths in the ocean."* Learning about the nature of the underwater light field went a long way toward illuminating the reasons behind the awesome strangeness of deep-sea eyes.

If you dive in a submersible to the very edge of darkness, you can see for yourself the profound changes in the visual vista that account for these eccentricities. This is where the scene transitions from an increasingly dim overhead light field to one that is defined by sparks of bioluminescence seen against pitch blackness. Many animals inhabiting this zone hunt by spotting the small silhouettes of prey swimming above them. Others detect flashes of bioluminescence directly below or in front. Some do both.

Take, for example, the deep-sea brownsnout spookfish (*Dolichopteryx longipes*). Its unfortunate common name notwithstanding,[†] this is an amazing fish, with a big head graced with *four* protruding eyes.[‡] Two large eyes point up and have prominent lenses that collect dim downwelling light, while two small eyes point down and employ mirrors to reflect and focus the light from bioluminescent point sources. The cockeyed squid (*Histioteuthis heteropsis*) meets the same challenge using only two eyes; one, the left, looks up and is large and bulging, while the other, the right, looks down and is small and sunken. A third solution is manifested in the Pacific barreleye fish (*Macropinna microstoma*). It can actually rotate its telescopic eyes *inside* its head! Although this exceptionally odd-looking fish has a black body, its head is transparent, creating a protective dome over the eyes, like the canopy on a fighter jet. It's no wonder early investigators were baffled by such outlandish adaptations.

There was also bewilderment at the relative eye size of animals inhabiting different depths. With eye size, there are two key factors

---

*John Murray and Johan Hjort, *The Depths of the Ocean* (London: Macmillan, 1912).

† Middle school must have been hell.

‡ High school was probably no picnic, either.

in play. Sensitivity is the first: Bigger eyes collect more light. Cost is the second: Larger eyes are energetically more costly to build and maintain.

Life demands energy, and energy in the deep ocean is generally in short supply. Most of the fuel for life comes from the sun. This is true despite the fact that sunlight bright enough to drive photosynthesis is found only in surface waters no deeper than about 650 feet. With the exception of the chemical energy found at hydrothermal vents and cold seeps, which provide a tiny fraction of the total available resources, the most common food source, even in the *sunless* depths, is derived from photosynthesis.

Plant life in the form of phytoplankton grows in surface waters, then either sinks into the depths as it dies or is carried there by consumers like jellies, crustaceans, squid, and fish that vertically migrate into deep waters and die or defecate, thereby releasing valuable foodstuffs. For deep-sea dwellers, it's all manna from heaven. But it is not an infinite bounty, and as the shower of food is gobbled up on its way to the bottom, the downpour turns into a trickle. It makes perfect sense, then, that the numbers and sizes of animals diminish the deeper we go.

What didn't make sense to early investigators, though, was this: As animal size *decreases* with depth, relative eye size *increases*, right up to the edge of darkness. Then the trend gets flipped upside down, because below that transition zone, eye size generally shrinks with increasing depth. Why?

Well, as I learned the hard way postsurgery, the real key to eyesight is not detecting light; it's being able to distinguish the brightness differences between an object and its surroundings—in other words, *contrast*. For animals that need to differentiate a small dark silhouette against a dim background light, the best way to enhance the contrast of that image is to gather more light from the background. This is better accomplished with a larger eye. Therefore, at deeper depths within the twilight zone, where the background light becomes dimmer, the eyes of upward-looking visual predators are generally larger. On the other hand, a bright object such as bioluminescence, seen against a black background, is essentially infinite contrast, which can be efficiently detected without resorting to the expense of a large eye.

Leonardo da Vinci said, "The noblest pleasure is the joy of understanding." To have a confusing set of facts and observations morph into a simple and elegant explanation produces a wonderful "aha" feeling. Understanding the nature of the light field and the challenges it presents for survival makes many things that once seemed inexplicable obvious.

For example, there was that silvery fish I saw hanging vertically in the water during my first dive in Wasp. It was a cutlass fish, so named because of its bladelike appearance and elongated body—as much as five feet long—which tapers to a point at the tail and is so shiny it looks like polished silver. I've seen many such fish during submersible dives and have often heard them described (by non–fish biologists) as "pogo fish," because they hover just above the seafloor and every time their tails make contact with the bottom they react by lunging upward. The effect can be bizarre. When there's a group of them together, it looks like a thicket of double-edged, hiltless swords hanging vertically in the water, doing an unsynchronized dance in which one after another darts straight up and then gently drifts down. It's wholly nonsensical, until you grasp the role that light plays in their lives.

Cutlass fish are voracious carnivores with fanglike teeth and large eyes that they use to hunt in the twilight zone, searching for silhouettes overhead. Assuming a vertical orientation facilitates their hunting strategy, permitting them to look straight up and at the same time present the smallest silhouette possible to any predators below them. They have a preferred light zone in which they hunt, so when we show up in submersibles or ROVs with our big, bright floodlights, they adjust their depth downward to get away from the light, until they hit bottom and can't anymore, causing them to lunge up-ward. Sometimes they swim away, but more often they hang around, stuck in a behavioral loop, performing their crazy pogo dance.*

How many other bizarre activities and adaptations might be ex-plained if we had a better understanding of the light field in the ocean—not just the solar light field, but the living light field? For example, there's the conundrum of eye size all the way down on the deep-sea floor. This is another transition zone where the general trend of eyes getting smaller as you go deeper below the edge of darkness undergoes a changeover. Many bottom dwellers have eyes that are large relative to their body size, and for those living at depths where sunlight doesn't penetrate, the logical explanation is that those eyes must be adapted for seeing the only light available: biolu-minescence. The trouble with this notion is that, unlike in the mid-water, where most of the animals are bioluminescent, the number of light emitters found on the ocean bottom is relatively low.

TO INVESTIGATE THIS puzzle, in 2009, Tammy Frank, Sönke Johnsen, and I put together a NOAA-funded expedition to explore bioluminescence on the deep-sea floor. We wondered if it was pos-sible that the assumed dearth of bioluminescence could simply be the result of nobody having seriously looked for it. To this day, most sampling of bottom dwellers depends on brutal forms of deep-sea trawling that leave the critters thoroughly thrashed. Perhaps they were luminescent but we didn't know it, simply because their light-producing chemicals were being exhausted during capture. Even samples collected by submersible or ROV might not survive getting to the surface because they are generally transported in such a way that they are not sufficiently protected from the dramatic change in temperature and so they are cooked before being examined for light production.

---

*A thankfully short-lived punk rock dance craze.

To test this notion, we proposed using the Johnson-Sea-Link (1) to carefully collect bottom dwellers by placing them in a thermally insulated box, called a BioBox, to keep them cold for the trip to the surface, and (2) to also literally poke around on the bottom with the sub's manipulator, while the lights were off, to see if we could stimulate anything to light up. The question was, where should we go hunting?

Three-quarters of the seafloor appears featureless and depauperate, while the other quarter more than makes up for it by including some of the most otherworldly communities on the planet. When I encountered one of these bewildering locales for the first time, it left me dumbstruck.

It was back in 1985, when I was diving Deep Rover in the Monterey Canyon. I was running with lights off for what was supposed to be a midwater bioluminescence transect. Ordinarily, readings from the sub's sonar would allow me to stay a safe distance above the seafloor. Unfortunately, my sonar was on the fritz, so I was depending on the surface ship to tell me how close I was, and they got it wrong. To be fair, they got it mostly right. I was tooling along for several minutes, thoroughly absorbed in the bioluminescent displays flashing all over my transect screen, when I plowed smack into the bottom. Apparently, during the second half of the transect, the seafloor started angling up, which nobody noticed.

Colliding with solid things in the dark while a couple thousand feet underwater is highly discomforting. The surge of adrenaline flooding my body as I flipped on the lights made me question the reality of what I was seeing.

I had blundered into an undersea garden straight out of the imagination of Dr. Seuss. Surrounded by enormous fans of rosy bubble-gum coral,* I was sitting right next to a giant yellow sponge shaped like a lacy, upside-down witch's hat. I could make some sense of the

* *Paragorgia arborea.*

coral and sponge, but covering the seafloor between them was a field of outsized mushrooms that looked like they belonged in a Pixar movie or a magical kingdom inhabited by unicorns. Some were white and some were pink, and each was adorned with long, feathery plumes poking out all over their mushroom caps. Sitting underneath one of these mushrooms, an orange frog looked up at me with such fixed intensity, I half expected him to scold me for disturbing his slumber.

I later learned that the mushrooms were a kind of soft coral* I'd never seen before—a relative of sea pens—and the "frog" was a large-eyed rockfish that was resting on its pectoral fins on the bottom, with its tail tucked behind it in such a way that it looked far more like frog than fish. Compared with the animals I was used to encountering in the midwater, these creatures were incongruously gigantic. Tellingly, they were all—except for the fish—detritivores.

Detritivores are so named because they survive on a diet of detritus, that is, the rain of food particles falling from above. Sponges constantly suck water through the tiny chambers that make up their body wall, where bits of foodstuff are filtered out before the current is expelled into the central barrel. The knobby-armed bubblegum coral and the mushroom-shaped sea pens feed by extending polyps into the water to grab passing particles. Since detritus is the primary food source, it's logical that the giants would show up where it accumulates at an interface like the bottom. Yet such giants do not occur everywhere on the seafloor, but only in places like the Monterey Canyon and on and around seamounts and rocky outcroppings where there is just the right combination of enhanced productivity in surface waters, a hard bottom to provide solid points of attachment, and particle-carrying currents.

In 2009, the place we picked to explore was yet another Seussian

---

*Mushroom coral, or *Heteropolypus ritteri*, formerly *Anthomastus ritteri*.

garden, even more fantastical than the one I slammed into in the Monterey Canyon. This one covered the slopes of a series of parallel limestone mounds in deep waters off the west end of Grand Bahama Island. Each of these mounds is shaped like an upside-down ship's hull, with sizes ranging from small dugout canoes to major cruise liners and oriented north-south, parallel to the Florida Current. Each was essentially an oasis surrounded by a sedimentary sea, allowing us to sample both types of seafloor habitats.

Anticipating that trip, I was of two minds about what we might expect to find in terms of the number of light emitters on the deep-sea floor. On the one hand, all those big eyes suggested a significant selective advantage for being able to see bioluminescence. On the other hand, since much of the bioluminescence in the midwater is related to playing hide-and-seek in a world without hiding places, it could also be argued that on the bottom, where there are structures to hide behind and no need for counterillumination to conceal silhouettes from upward-looking predators, bioluminescence might be much less prevalent. This is certainly the case in the coastal zone environment, where only about 1 to 2 percent of bottom-dwelling species are bioluminescent, as compared with more than 75 percent of midwater species, but in the coastal zone there is also lots of illumination from sunlight and moonlight, making bioluminescence nonessential for visual communication. The only way to find out for sure was to go look for ourselves.

This whole expedition was a visual feast, which began with my first dive on the site. As we descended at the north end of a large mound, it appeared, like a steep-sloped island encircled by a white sand ocean. As we headed upslope toward the "keel" of the mound, we encountered a fantastic-looking landscape of row after row of straw-colored stalked crinoids. More commonly known as sea lilies, these ancient creatures look more like miniature cartoon palm trees than the sea urchins and sea stars that are their closest living relatives. The mouth is located at the center of their crown of feathery palm fronds, which are their arms. Each mouth was pointing toward us as we proceeded upslope against the current, while their arms curved backward into the current, slowing the flow and making it easier for the tube feet to capture particles from the water.

There was little to no space between individuals down the length of each row, but there were large gaps between the terraced rows that were oriented perpendicular to the current, which appeared to optimize detritus collection on the down-current side of the mound. One of our collaborators on this expedition, Chuck Messing, was a leading authority on crinoids. We knew he'd be thrilled by this find, and we collected specimens for him to identify and for us to test for bioluminescence. Bioluminescence is rare in crinoids, but not unheard of, so I was disappointed when it turned out that these and all the other crinoids we examined during this expedition proved to be non-luminescent.

As we proceeded south along the top of the mound, a very different vista loomed into view—large, fanlike "trees" of magnificent golden coral* over three feet tall and six feet wide and ranging in color from canary yellow to rust orange to auburn brown. We knew these were very likely to be bioluminescent (as similar species are), and sure enough, when we turned out the lights and brushed the sub's robotic arm through their soft branches, they glowed an intense turquoise. Smaller delicate sea fans were also present, forming a dense understory. Most were gold-colored, but a few were a vivid purple. These all proved to be non-luminescent.

Colonies of white stony coral,† also non-luminescent, studded the hillside at random intervals, their fragile branches poking out at disparate angles, making them appear disorganized compared with the delicate flattened tracery of the fans, with their intricately spaced, nonoverlapping branches always oriented perpendicular to the cur-

---

* *Parazoanthidae.*
† *Lophelia.*

rent. There were also feathery fronds of bamboo coral,* so named because their internal skeleton is made up of alternating long white and short black bands, resembling, in shape, a bamboo stalk. All of these proved to be luminescent. In some species, the coral branches glowed neon blue anyplace they were touched, while others produced twinkling displays that looked like a profusion of miniature cerulean-blue Christmas tree lights winking on and off.

Although they look like trees and bushes, all these fantastic life-forms are animals—specifically, detritivores. And perched in their branches were more detritivores: brittle stars, snake stars, sea stars, gooseneck barnacles, hydroids, sponges, anemones, feather stars, and basket stars. We collected samples of all of these and more, but in the end, fewer than 20 percent of all the animals we tested for bioluminescence proved capable of making light.

Most of those light makers were already known to be luminescent, but a few were new and wonderfully strange. My favorites were the anemones. There were Venus flytrap anemones,† which look like a bright orange version of Audrey II from the musical *Little Shop of Horrors,* but instead of sucking blood, when prodded they squirt out strings and swirls of sticky cobalt-blue luminescence. This probably deters predators that don't want to make themselves targets for their own visual predators.

Even more wonderful were the two bioluminescent anemones that we found attached to the seashell home of a hermit crab. These glowed when stroked and brought to mind wild imaginings of a deep-sea existence toting around a mobile home adorned with elaborate Victorian lamps that light up only when rubbed like Aladdin's lamp.

These were wonderful discoveries, but they paled in comparison with what we witnessed when we turned out the lights and sat qui-

---

* *Alcyonacea.*

† *Actinoscyphia.*

etly in the dark. Whenever I did this in the midwater, as long as I went dead still I saw nothing—no spontaneous bioluminescence, just absolute and complete blackness. But here, on the seafloor, there was frequent luminescence, not from the detritivores living on the bottom but from plankton carried by the current, which were mechanically stimulated when they bumped into the detritivores. I recorded some video of this with my intensified black-and-white camera, which showed brief flashes of light in the branches of the golden coral. But the best imagery came from Sönke, who used his Nikon to take a ten-second color exposure. Just before capturing the image, he had the pilot sweep the manipulator through the golden coral, causing it to glow. In the image, you can clearly distinguish the coral branches studded with distinct polyps glowing blue-green, while the plankton hitting it and passing through appear as blue streaks.

We measured the emission spectra of all the animals we collected and found that many of the bottom dwellers produced a greenish light, rather than the blue that dominates in midwater inhabitants. Suspended sediment near the bottom favors green-light transmission over blue, so the color shift may reflect selection for maximum visibility. A similar shift toward greener emissions is seen in some bioluminescent inhabitants of sediment-laden coastal waters.[*]

We wondered if the bioluminescent color difference between the detritivores and the plankton that bump into them could explain another intriguing mystery that had emerged from Tammy's research. During our Deep Scope 2005 mission, she had discovered that a squat lobster called *Gastroptychus spinifer* appeared to have two different color receptors in its eyes. With results from only one animal, she didn't feel she could publish such an unusual finding, but on

---

[*] That oh-so-valuable molecule green fluorescent protein (GFP), which has been used to greatly advance cell biology research by illuminating the inner workings of cells, was extracted from the jellyfish *Aequorea victoria* and is believed to be an adaptation to shift blue bioluminescence to green in this coastal species.

this 2009 mission she managed to get more specimens that confirmed it. To see color requires giving up sensitivity, so we were very curious about what this extraordinary adaptation might be for.

Although the flattened reddish-orange bodies of these squat lobsters make them look more like lobsters than the hermit crabs to which they are most closely related, their lifestyle and feeding strategy set them apart from either group. We saw many of these guys perched high in the branches of the golden corals, with their absurdly long arms spread to either side and their pincer-like claws splayed open. Their large eyes are on stalks that are not on either side of the head, but side by side and forward-facing in a way that would potentially allow binocular vision—in other words, depth perception.

Based on Tammy's results and our observations of the nature of luminescence on the deep-sea floor, we speculated that they might use color vision as a way to distinguish the blue luminescence of plankton hitting the coral from the blue-green luminescence of the coral itself, their pincer claws and binocular color vision allowing them to pluck food directly out of their hosts' mouths—making them less-than-delightful houseguests.

Being able to observe the extraordinarily alien visual environment of the deep sea is essential to understanding the life therein. Although the number of luminescent species we found on the deep-sea floor was much fewer than what we discovered in the midwater, the amount of spontaneous bioluminescence was vastly greater. It seemed very likely that food in the form of bioluminescent plankton that lights up when it encounters the deep-sea floor must be a valuable clue for many big-eyed seafloor inhabitants.

THERE IS STILL another vital food source that may light up when it encounters the deep-sea floor, and that is marine snow. The term "marine snow" was first coined by William Beebe to describe the daily slow-motion sprinkling of food into the deep sea. As usual, he nailed it. It looks very much like snow—white flocculent particles resembling everything from a slow-motion flurry to a blizzard. But if you look closer, you notice differences—single flecks, white fluffy bits, messy clumps. The Inuit supposedly have more than fifty words to describe the various forms of snow, and marine snow may be deserving of similar linguistic largesse.

Surprisingly, this diversity is sometimes visible even with the lights off. Obviously, this is possible only when the "snow" is luminescent, but in my experience a lot of it is. Bioluminescence from marine snow must be stimulated either mechanically or with light, and once stimulated it does not readily reignite, which makes sampling it a challenge.

The best way to see it is to either ascend or descend through the water column with the lights out and intermittently flash a light into the darkness. If you use a flashlight, you get a localized response. (One of the sub pilots told me about a time when the light-stimulated marine snow was so dense that he could write his name in light.) But you get a more spectacular response if you flick the sub's floodlights on and off. When you do, as soon as your artificial light is extinguished, you are surrounded by a snowstorm where all the snowflakes simultaneously flash on and then gradually fade out. But it's not winter snow—it's marine snow, like the mermaid's tears I observed on my first dive in Wasp, short strings of tiny glowing orbs encased in a wispy sheath, or fragile gossamer aggregates that look like miniature bottlebrush flowers with specks of light at the end of each filament, or some other equally fragile configuration. If you turn your lights on, the source of the light will likely be invisible. You may see flecks of marine snow, but whether they were the cause of the light is not obvious.

Marine snow is very, very important. It is the primary source of food in the deep, and therefore it seems like we should know a bit more than we do about its bioluminescence and what role it might play in the survival strategies of deep-sea fauna, but it has been an

incredibly difficult phenomenon to document. With recent developments in camera technology, I have real hope that revelations may
come soon—hopefully before I shuffle off this mortal coil—because
it's one of those profound mysteries at the edge of the map that not
only is intensely intriguing but may also prove extremely significant.

Specifically, I would like to understand the role of marine snow
in what is known as the biological pump, the carbon cycle in the
ocean, which is of more than passing interest these days because of
the part it plays in lowering atmospheric carbon dioxide, and thus
slowing global warming.

Years of observing it firsthand have led me to believe that most of
the bioluminescence I have seen associated with deep-sea marine
snow is bacterial. What makes this a somewhat controversial idea is
the fact that the bioluminescence needs to be stimulated. Bacterial
bioluminescence is very different from most other kinds of bioluminescence, because instead of producing flashes of light, it emits a
persistent glow. This is because its light chemistry is directly linked
to its breathing chemistry (better known as the respiratory chain).

Most people are familiar with bacterial bioluminescence as it relates to animals like anglerfish and flashlight fish. These species
don't manufacture their own light-producing chemicals but instead
co-opt the light produced by bacteria and, in return, provide the microbes with food and shelter within growth chambers. In the anglerfish, the chamber is called an esca and is located at the end of a long
fin ray that serves as a fishing rod, dangling the glowing morsel
within reach of the angler's toothy maw. In the aptly named flashlight fish, the bacteria reside in a large light organ just below the eye.
To turn off its flashlight, this fish actually has the equivalent of an
eyelid that closes up, shuttering the shine.*

Fish and squid that have evolved symbiotic relationships with

---

* In some species, the light organ actually rotates backward like the headlights on
your Lamborghini.

bacteria control the light in a variety of ways, usually through mechanical shuttering, but in some cases it's thought that the light is
modulated by controlling the availability of oxygen, because without
oxygen, bacteria don't glow. The Exploratorium, in San Francisco,
has a lovely demonstration of this in its exhibit of bioluminescent
bacteria. When I first saw it many years ago, they were maintaining
cultures of bioluminescent bacteria in flasks held on a shaker table.*
When the table wasn't shaking, no light came from the flasks, but
when the shaker was activated, the culture was stirred, introducing
oxygen that activated the glow. Recently, they came up with a more
elaborate scheme to demonstrate the same phenomenon. The bacterial broth is held in a thin tank with air inlet ports that visitors can
control to create swirling patterns of living light.

The point is that oxygen must be present for the bacteria to glow.
In marine snow, phytoplankton and other organic matter are broken
down by microbes that consume oxygen. This is a portion of the biological pump that releases carbon dioxide into the water. Although
the marine snow may be surrounded by oxygen-laden water, within
the particle itself there is an anoxic microenvironment. Bumping the
snow floc is just like activating that shaker table, introducing oxygen
and allowing the bacteria to glow. Shining light on the snow may
also introduce oxygen by stimulating its production in photosyn-
thetic bacteria, known as cyanobacteria, which are also commonly
found in marine snow. That might explain *how* marine snow
glows . . . but it doesn't explain *why*.

When I was starting out in the field of bioluminescence, one of
the hotly contested issues of the day was how bioluminescence could
have ever first evolved in bacteria. In other words, how could producing light possibly help a single bacterium survive? This was especially vexing from an evolutionary perspective because a single

---

*Not a table designed by Shakers, but an oscillating platform used to stir sub-
stances in a tube or flask.

bacterium doesn't produce enough light to be visible to any known eye; its emission is simply too dim. The only way bacteria are visible is if there are millions of them together, so how could the first bacterium to emit light have been selected for? What is even more confounding is that if you mix together two strains of the same species, one luminescent and the other dark, the dark mutants rapidly overrun the culture, because the energy cost of producing light puts the light emitters at a disadvantage. Any way you look at it, it seems like the deck was stacked against the evolution of bacterial bioluminescence.

A possible solution to this conundrum was revealed when some Polish scientists tried irradiating a mixed culture of dark and light-emitting mutants with ultraviolet light. When they did, suddenly the situation was reversed: The light emitters seemed to have the advantage and took over the culture. The bioluminescent strain continued to predominate as long as the UV light was on, but as soon as it was turned off, the dark strain regained supremacy. The reason was related to the fact that UV light damages DNA, which is why you should always remember to wear sunscreen.*

A UV photon packs a significant energy punch—more than blue or green or any of the colors of the rainbow—enough to mess with DNA's structure. Because UV light is so damaging, bacteria evolved a remarkable enzyme called photolyase, which repairs UV-damaged DNA. Intriguingly, this enzyme requires visible light—blue light, in fact—to work its magic, so apparently the selective advantage of bioluminescence is that it can stimulate DNA repair,† even at light levels that are too dim to be visible.

The need for cell repair mechanisms is now thought to be an

---

* Preferably sunscreen without the coral-damaging chemicals oxybenzone and octinoxate.

† You can actually purchase these powered-by-light enzymes in some "age-defying" skin care products . . . for only half the price of beluga caviar.

underlying basis for the evolution of a lot of different bioluminescent chemistries. One of the many challenges that life on Earth must deal with is not just the destructive properties of UV light but also the damaging effects of oxygen. We think of oxygen as a good thing, because without it we would die, but it does have a downside. It's so hungry for electrons, it will rip them from vital molecules like DNA and proteins. It is because of this destructive effect of oxygen that consuming antioxidant-containing foods like fresh fruits and vegetables is so important. Antioxidants prevent cell damage of a kind that is linked with aging and ailments like cancer, Parkinson's, Alzheimer's, and heart disease.

As it turns out, a lot of luciferins are endowed with antioxidant properties. In many cases, luciferins first evolved as scavengers of oxidants that are toxic to cells. Only later, with the evolution of specific luciferases, did these molecules become involved in light production. Interestingly, in bacterial bioluminescence, it appears that the enzyme rather than the substrate is the detoxifying agent, but in both cases the underlying principle is the same: Key elements of the bioluminescent chemistries originally evolved because they provided protection from oxidation.

In the case of bacteria, based on the Polish experiment described above, it seems that the next key evolutionary step involved producing dim light to protect against cell damage from UV light. For that light to then become bright enough to be visible required a whole different kind of selective advantage—one that may help ensure a reliable food source for the bacteria in a food-poor environment.

Although it has been said that you can't polish a turd,[*] in the ocean you can. In fact, if you're the right kind of bacteria, you can make it positively *shine.* My first acquaintance with this phenomenon was as a graduate student in Jim Case's lab. My classmate Mike

---

[*] Often used in reference to junk cars—like the VW bug that David and I rebuilt out of spare parts.

Latz was involved with an experiment that demonstrated the camouflage trick of counterillumination in a deep-sea shrimp. During the course of the experiment, the shrimp, which was being held in a specially designed light-measurement chamber, rather suddenly began putting out a very bright and prolonged light emission that was in no way related to the overhead light it was trying to match. Upon inspection, it was discovered that the shrimp had produced a brightly glowing fecal pellet. What I love about this story is that the record of that "event" actually got published in an article about counterillumination in the prestigious journal *Science*. In the context of the article, there was no good reason to include the event, except that I think it appealed to Dr. Case's perverse sense of humor.

It turns out that a lot of marine fecal pellets glow because the bacteria that help decompose them are bioluminescent. The reason for this, according to the "bait hypothesis," is that by glowing en masse (*dans la merde*), the bacteria make themselves easy targets for visual consumers. The bacteria are consumed along with the pellet, thereby gaining access to the food-rich environment of the consumer's gut. This now puts them at a significant advantage over non-luminescent bacteria, because dark bacteria will be essentially invisible and therefore far more likely to sink into the depths, where food sources are limited. This was a pretty remarkable insight that Bruce Robison and others first put forth in an abstract* published way back in 1977 and others have expanded on since.

As pellets sink into the deep sea and away from the harmful UV radiation of sunlight, the selective advantage of bioluminescence for the part it plays in DNA repair vanishes. When an energy-demanding process is no longer useful, it usually disappears, because mutations that diminish that process will be selected for. However, in this case, another selective advantage came into play because it favored

---

*A scientific abstract is a short overview of a scientific paper or presentation. This was a summary of a presentation made at the Western Society of Naturalists.

bioluminescence for a wholly different purpose: providing better access to nutrients. By glowing, the bacteria attract roving food buffets, which is to say, the stomach contents of their consumers, but this strategy works only if there are enough bacteria present to be visible.

And so another amazing adaptation evolved, called quorum sensing. This remarkable trick allows bacteria to communicate with one another and thereby coordinate their efforts to benefit their mutual survival. In the case of bioluminescence, it assures that the bacteria do not expend energy manufacturing their light-producing chemicals unless they are present in sufficient numbers to be visible. In order to get the equivalent of a head count, the cells produce a small signal molecule. When the concentration of this molecule reaches a certain threshold level, it triggers a change in gene expression and the cells start producing the chemicals they need to emit light. Although quorum sensing was first discovered in bioluminescent bacteria, later this method of communication among bacteria was found in a surprising variety of bacterial processes, including virulence, antibiotic production, and motility.

GLOWING POOP IS a concept with a lot of appeal.* For instance, it helps explain why the escas on anglerfish function as lures: because they mimic a common food source, morsels of glowing excrement. And there is good circumstantial evidence for the bait hypothesis. For example, gut content analyses of fish have revealed an abundance of luminous bacteria; it has been confirmed that luminous bacteria survive passage through fish guts; and laboratory experiments have demonstrated that zooplankton inoculated with glowing bacteria are more readily located and preyed upon by nocturnal fish.

If populating fecal pellets benefits the survival of bioluminescent

---

*Just imagine the possibilities in the novelty item industry alone.

bacteria, then it seems to me that populating marine snow might be similarly beneficial. But based on my observations, "snow shine" requires either mechanical or photic stimulation. I have sat in submersibles for long stretches with the lights out without seeing a hint of snow shine, until I flicked the submersible's floodlights on and off.* When I did, the flashback could be anything from "Meh" to "Holy smokes!" but there's always something. And while the shapes may vary, the kinetics are always the same. It ramps up to full brightness quickly, glows for many seconds, and then slowly fades to black. The first such response is always the best; subsequent flashbacks are generally dimmer.

Perhaps this requirement for stimulation serves as a means of energy conservation. You know the old question "If a tree falls in the forest and no one is around to hear it, does it make a sound?" When marine snow falls into the deep sea and there's no one there to see it, does it glow? My guess is no. It needs to be stimulated, either mechanically or by light—perhaps from the bioluminescent flashlights displayed by so many deep-sea denizens. Unless there is someone there to see and consume it, the snow would remain dark until it encountered the bottom, resulting in the mechanical stimulus that would cause it to light up. Evidence that this may be the case has come to light from a very unexpected source: physics.

Shortly after I started my graduate studies with Jim Case, I answered the phone one day in the lab and found myself speaking with an overwrought physicist. He was part of a major neutrino detection† project that had placed a large array of ultrasensitive light detectors deep in the ocean, as far as possible out of the reach of sunlight. Their idea was to find the darkest place they could, and

---

*Two short pulses of light work better than one.

†Neutrinos are created by radioactive decay, including that which occurs in the core of a star, making their detection useful for a branch of astronomy known as neutrino astronomy, which allows astronomers to explore the cosmos with new eyes.

these detectors were intended to identify neutrinos by the dim flashes of light generated as charged particles streak through the water faster than the speed of light.*

The trouble was that their detectors were seeing way more light than they should. The physicist called our lab because someone had suggested that the light might be bioluminescence. In a voice that was actually shaking, he asked, "Can this be true?" I assured him that it could. There was a long pause and then he asked, "Is there some place in the ocean where there isn't any bioluminescence?" My answer—"Not that I know of"—did not make this man happy.

It may seem incomprehensible that this large, expensive project got funded with such a major flaw in the experimental design, but that is simply an indication of how little known the full extent of bioluminescence in the oceans was—and still is! That project, known as DUMAND (Deep Underwater Muon and Neutrino Detection), struggled for nearly two decades between 1976 and 1995, suffering an endless string of technical difficulties associated with trying to install an array of ultrasensitive light detectors 15,750 feet deep in the Pacific Ocean off the Big Island of Hawaii, before it was finally abandoned.

Another project with similar goals has since replaced it. Situated 8,200 feet deep in the Mediterranean Sea off the coast of France, this neutrino detector goes by the painfully contrived acronym Astronomy with a Neutrino Telescope and Abyss Environmental Research project, better known as the ANTARES telescope. This system, too, has had issues with bioluminescence that has impacted

---

*Although nothing can travel faster than the speed of light *through a vacuum*, some particles can travel faster than the speed of light through water, and when they do, they generate light called Cherenkov radiation, named after the Russian scientist who first demonstrated it. Just as an aircraft traveling faster than the speed of sound through air emits a sonic boom, a particle traveling faster than the speed of light through water emits a kind of light boom. Cherenkov radiation is the cause of the blue glow associated with underwater nuclear reactors.

its detection limits and required some sophisticated background suppression tricks, but not only does it work for detecting neutrinos, it has also resulted in the longest continuous time series of deep-sea bioluminescence ever recorded.

Several noteworthy observations have arisen as a result. The first is that, in an effort to better understand the bioluminescence they were observing, they did some sampling that provides—at minimum—circumstantial evidence that most of the bioluminescence they have recorded with the ANTARES telescope is the result of bioluminescent bacteria and that these bacteria are associated with particles rather than being free-living. The researchers also did tests on the impact of pressure on some of those bioluminescent bacteria and discovered that they emitted five times more light at high pressure than at low pressure—which suggests that they are uniquely adapted for life in the deep sea. And most intriguing of all, they found a linkage between seasonal periods of intense marine snow blizzards and increased bioluminescence recorded by the telescope.

The fact that a marine snow blizzard hitting the deep-sea floor may trigger a lot of bioluminescence could help explain how animals can visually locate food even in the three-quarters of the sea-floor that appears devoid of life.

BESIDES MARINE SNOW, another major source of food that accumulates on the deep-sea floor arrives in the form of dead creatures. These food packages are sometimes called deadfall or, for the largest of them, whale fall, and their size makes them a bonanza, for which there is stiff competition. Here, too, bioluminescence must come into play, as large food falls may be revealed not just by downstream scent trails, which are often highly directional, but by bioluminescence, which can be omnidirectional and visible over considerable distances.

The arrival of a food fall on the abyssal plane could be revealed by the mechanical stimulation of bioluminescent marine snow and plankton colliding with a carcass that protrudes into the bottom current. Or bioluminescent animals attracted to the bait may be stimulated to luminesce to defend themselves from other onsite predators. Or, if the bait becomes infected with bioluminescent bacteria, it may glow like a supersized fecal pellet. In any of these cases, large mobile predators with eyes, like giant sixgill sharks, would be at a distinct advantage. First come, first served.

The prevailing thinking on gigantism like that of sixgill sharks is that their size is advantageous because it permits greater energy storage, allowing long swims between food buffets. There has been some concern recently among deep-sea biologists that those swims must be growing longer thanks to the extreme impact of overfishing, which has greatly reduced the number of food falls reaching the deep-sea floor. However, there was still one more amazing finding that came out of our Deep Scope missions, which suggests that sixgill sharks have a feeding trick that may afford them an alternative energy source.

During our Deep Scope mission to the Bahamas in 2007, we saw sixgill sharks that were apparently drawn into the vicinity of the EITS camera either by the scent of the bait or by the visual stimulus of the e-jelly, but instead of going after the bait, these mammoth creatures oriented their bodies vertically in the water with their heads down as they sucked up sediment, then blew it out of their gills in billowing clouds. Our presumption was that they were feeding by sieving out burrowing creatures that inhabit the upper layers of soft sediment.

Although the vast, featureless regions of the deep-sea floor were once thought of as desertlike, we now know this not to be the case. In fact, an impressive assortment of small creatures like worms, crustaceans, gastropods, and nematodes burrow through and feed off the detritus that settles there and might well serve as a resource for these giants when food falls are scarce.

But before we could publish this remarkable finding, we needed to do a follow-up study, in which we planned to collect sediment samples at each deployment site so we could document that this was indeed a form of feeding. Unfortunately, we never got the chance. The 2009 expedition was my last mission with the Johnson-Sea-Link submersibles. In 2010 Harbor Branch retired them, sold off the last of its ships, and shortly thereafter laid off the crews. It seemed that the golden age of submersibles was coming to an end.

Chapter 13

THE KRAKEN REVEALED

*K*ABOOM! It was a deafening crack, not something you want to hear on a ship far from shore. The lights went out, and I bolted for the fantail along with the other scientists who had been crowded around the laptop excitedly viewing the video we had just downloaded. As soon as we stepped on deck, it was clear the ship had been hit by lightning. Bits of the ship's antenna were all over the deck, and a pillar of yellow and brown smoke was drifting aft. *Never in all our considerable cumulative seagoing experience had any of us been on a ship that was hit by lightning.* As we were remarking on that fact, we all seemed to have the same thought at the same moment: *Crap! We didn't back up the video! Did the computer get zapped?* To have possibly lost the first footage ever recorded of a live giant squid in U.S. waters was a horrifying thought.

Among marine biologists, the giant squid often serves as a symbol, much like Captain Ahab's great white whale, for "the one that got away." Jokes and references to giant squid are just part of the seagoing culture. The tension on a trawl line suddenly increases and somebody will say, "Must have caught a giant squid." A net comes up empty and shredded and the giant squid gets blamed. To claim to have filmed a giant squid and then have to say "You'll just have to take our word for it" would never cut it.

For some marine biologists who have spent their careers hunting the giant squid with Ahab's fervor, the opportunity to be the first to see the world's most famous invertebrate in its natural habitat was the goal of a lifetime. For others, like me, it was merely a highly improbable fantasy.

ANCIENT MARINERS TOLD many tales of terrifying sea monsters whose size and ferocity grew with every flagon of spirits consumed during the storytelling. One of the most famous of these leviathans was known to Norwegians as the Kraken, a titan that struck terror in the hearts of seafarers. They described it as a multiarmed beast so enormous it could be mistaken for an island when seen floating at the surface, and so deadly it could drag men and ships to watery graves. We now recognize their accounts as providing a fair description of what we know as the giant squid, which goes by the scientific name *Architeuthis*.

There was much scientific skepticism of these early accounts, but proof finally came in 1861 when a French warship operating off the Canary Islands happened upon one of these behemoths. It was apparently dying, but, not wishing to take any chances, the sailors fired some shots to finish it off before using a rope to try to haul it on board. The enormous weight of the beast caused the line to slice through its body, with the result that the multiarmed head end fell back into the sea.

Still, they made a sketch of what it looked like and managed to retrieve the tail end to corroborate their account. It was enough evidence for a paper on their observations to be presented at the French Academy of Sciences. The renowned writer Jules Verne saw this report and incorporated it into the novel he was then penning, *20,000 Leagues Under the Sea*, describing a bloodcurdling battle with the Kraken that only served to heighten its legendary fearsomeness.

No science-fiction writer could ask for a more fantastic alien to terrorize readers. Besides having eight muscular arms and two insanely long tentacles—all appearing to grow straight out of a massive cone-shaped head—this monstrosity was also equipped with a parrotlike beak to rip flesh, serrated suckers to perforate and lock on to the slimiest of prey, a jet propulsion system that works equally well in forward or reverse, three hearts that pump blue blood, and gargantuan eyes the size of a human head—bigger than those of any other animal on Earth.

Our fascination with big stuff sets in at an early age. Big dinosaurs, big machines, and big sharks, among so many other things, become obsessions for some kids. Perhaps it's a natural response to a time in life when your imagination is most fertile and almost everyone is bigger than you.* Marine scientists are intrigued by giants, too, not just because they are cool, which they definitely are, but because they are so at odds with most of the life that inhabits the depths. When outsized life-forms appear, an obvious question is how do they manage to find enough nourishment to grow so big in a place where food is so scarce?

Deep-sea gigantism is manifest in a fantastically odd assortment of creatures. There are crustaceans like the giant isopod, a pill bug the size of a Tonka truck, and the Japanese spider crab, which has a claw-to-claw leg span of over twelve feet. There's a seven-arm octopus that is as long as a Volkswagen Beetle. (Only the females grow that big. The males are much smaller but compensate with an unusual sexual modification that gives the species its name: One of its eight arms is specially modified for sex and hidden from view, neatly coiled in a sac beneath the right eye.†) There are also giant sharks like the Greenland shark that can grow up to twenty-four feet long; the giant oarfish, which is the world's longest bony fish, reaching a maximum recorded length of twenty-six feet; the giant squid, which

---

*Mammoth fantasy figures can inflict sweet revenge on big bullies, as exemplified by Calvin's *Tyrannosaurus rex* alter ego in the comic strip *Calvin and Hobbes*, by Bill Watterson.

†According to the Manly Man's Guide to Fashion for Manly Men, this is the latest trend, superseding color-coordinated pocket squares.

reaches lengths of at least forty-three feet; and lots of giant jellies, including a siphonophore discovered in a submarine canyon off the coast of Australia that is believed to be the longest ocean creature ever recorded, at 150 feet long.

Some giants live on or near the bottom, where food raining down from above accumulates, producing a greatly enriched nutritional environment compared with the open ocean. The isopod and the spider crab, for example, are scavengers that feed on carcasses and organic matter landing on the seafloor. When food is plentiful, they thrive and grow. When food is scarce, their metabolism allows for long periods of starvation. To be able to go without eating for not just weeks but in some cases months provides a major advantage in a food-poor existence.*

The Greenland shark adopts a similar strategy, cruising slowly above the bottom, consuming both live fish and animal remains, including seals, moose, and reindeer.† If you never stop growing, then the longer you live, the bigger you get. The Japanese spider crab can live as long as a century and, based on carbon dating, it's believed that Greenland sharks may live for up to four centuries. So it's not just their size that astounds; it's their longevity, because that kind of life span suggests there are titans cruising the depths now that were alive when the Pilgrims sailed on the *Mayflower*!

But what about the giants that inhabit the deep midwater, where the primary food is fecal pellets and marine snow? Here the food content is so dilute, it's the equivalent of a few grains of rice in a cubic meter (264 gallons) of seawater.‡ To survive on such a diet,

---

*One giant isopod in a Japanese aquarium supposedly went on living while refusing solid food for five years!

†As far as we know, the Greenland shark has no relation to the Landshark of *Saturday Night Live,* so presumably the moose and reindeer fall through the ice.

‡Compare the calorie count of one cubic meter of seawater containing ten grains of rice (about one calorie) with a cubic meter of beer, which would be seventeen kegs, or about 421,000 calories!

animals have to sieve through an enormous quantity of liquid—
anywhere from one hundred thousand to ten million times their
own body volume per day! Many smaller animals, like copepods and
krill, accomplish this by setting up a feeding current—they beat
their appendages to draw water toward their mouths, where they
scan for particles. But larger animals need to compensate for the fact
that feeding efficiency decreases with increasing size. One way to do
so is to greatly expand their capture volume by deploying huge
mucus feeding webs or other food-gathering contrivances.

Mucus is the duct tape of the ocean—it holds the marine universe together and can be used for all manner of strange constructs.
My favorite are the insanely elaborate mucus "houses" constructed
by tadpole-like creatures called larvaceans. A larvacean a couple of
inches long can construct a mucilaginous McMansion that is more
than three feet across. The most distinctive architectural feature of
the "house" is its filters, which look like the two lobes of a brain, with
parallel crests and furrows resembling a white lace Elizabethan neck
ruff. By beating its muscular tail, the larvacean forces water through
these food-concentrating filters. When the filters become clogged
with critters too large for the larvacean to consume, the house is
discarded. Some species produce as many as forty houses per individual per day.* Discarded houses and feeding webs then become a
food source for larger creatures, like the seven-armed octopus, that
are willing to tolerate a lot of mucus in their diets.

Jellyfish and their ilk, a group known as gelatinous zooplankton,†
have bodies composed of more than 95 percent water. Because seawater surrounds and supports them and they live where there is no

---

* Larvacean tips for successful house flipping: (1) low-cost construction materials,
(2) complete remodel in less than an hour, (3) DIY decorating with bioluminescent granules installed in species-specific patterns in house walls.

† A disparate group of soft-bodied, usually transparent animals that includes jellyfish, comb jellies, siphonophores, salps, larvaceans, and some worms and mollusks.

turbulence, they can expand their feeding volume to an enormous extent, with all kinds of fragile structures that would be impossible in air. The closest facsimile to such constructions on land would be a spider's web. A 150-foot-long siphonophore, for example, can deploy a deadly curtain of stinging tentacles that "fish" for prey such as small crustaceans and fish. Stinging cells called nematocysts immobilize and kill their victims as the tentacle pulls the meal into one of its many stomachs for digestion. It's a food-gathering strategy ideally suited to the midwater. Jelly giants appear in an impressive variety of forms, as do their predators, which include leatherback turtles, the largest of all living turtles, which can grow to over seven feet long and weigh more than 1,500 pounds; ocean sunfish, which can reach weights over 2,200 pounds; and our friend the seven-armed octopus, which finds jellies as yummy as mucus.

Even among this menagerie of oddities, the giant squid stands apart as an outlier. First of all, there's the question of its age. How long *does* it take an animal to grow to the height of a four-story building? As a general rule, squid have short life spans—three to five years—and exhibit rapid growth. Based on counts of what are thought to be daily growth rings in giant squid statoliths (balance organs equivalent to the human inner ear), it is speculated by some that giant squid can reach adult size in approximately one and a half years. That's equivalent to *doubling* in size every two and a half weeks.

On the other hand, rather than laying down growth rings daily, as seems to occur in most squid, it's possible that *Architeuthis* lays down a growth ring after every feeding episode. Carbon dating of statoliths suggests a life span of fourteen years or less,* which makes its rate of growth more plausible, but still impressive in a food-poor

---

* But this depends on assumptions about depth range and temperature exposure that leave the actual number very much in doubt.

environment. This is especially true because the giant squid suffers from an evolutionary quirk that precludes gorging.

Evolution is full of examples of trade-offs that have led to some fundamental shortcomings, a consequence of incremental improvements on preexisting systems rather than a holistic design. For example, the reason thousands of people die every year from choking on food is that we use the same passageway for eating as we use for respiration.* Squid have a very different kind of choking hazard on account of their gullet passing through a small hole in their donut-shaped brain. Consequently, if they try to swallow too large a bite, the result could be brain damage!

Stomach contents of dead giant squid indicate a diet that depends on fish and other squid. These represent concentrated food sources, but if energy must be expended to hunt them down and if, once captured, they can only be nibbled rather than gulped, it raises a lot of questions about the energy cost and return for food acquisition. As a result, there has been much debate about whether *Architeuthis* is an active predator that hunts prey or a sluggish sit-and-wait predator that conserves energy by drifting passively.

One of the giant squid's most notable features is its eyes, especially when you compare them with the eyes of its primary predators, sperm whales. Seen up close, a sperm whale's eye looks impressive: It's as big as a billiard ball. By contrast, the eye of a giant squid is five times larger—bigger than a basketball! Given the metabolic costs of growing and sustaining such a sensory organ, it's clear that vision must play a vital role in the life of *Architeuthis,* one that presumably involves detecting bioluminescence for either locating prey or avoiding predators, or both.

One hypothesis is that such large eyes help giant squid escape

---

*Which is why you should take small bites, chew your food slowly, and learn what to do in a choking emergency. No kidding!

capture by sperm whales. Although scars from giant squid suckers found on sperm whale skin indicate that the squid put up a fight, the number of giant squid beaks discovered in sperm whale guts suggests that the balance of power resides firmly with the whales.

While the toothed whales use reflected sound to locate their prey, the squid are thought to cheat death by using their enormous eyes to detect the bow wave of light created by whales swimming through a bioluminescent minefield. In that context, the enormous eye size makes sense, as the enhanced sensitivity it provides is directly linked to survival if it gives enough warning of an attack to allow time for evasive maneuvers.

Presumably, their large eyes might be similarly valuable for detecting the movements of large prey and homing in on bioluminescent burglar alarms. But how could we ever know for sure without direct observation?

GIVEN ITS ENORMOUS size and peculiarities, it is little wonder that such a fantastic, mysterious creature as the giant squid has been sought for so long. Getting footage of it in its natural habitat became the holy grail of natural history cinematography, and there were many efforts, including two major multinational expeditions off New Zealand in 1997 and 1999. But all attempts failed, resulting in several documentaries that ended with the chief scientist standing at the bow of the ship watching the sunset as the narrator intoned a moving tribute to the tribulations of exploration.

Because of the prohibitive costs involved, there had been no serious financial backing available to try again until 2004, when Japanese squid scientist Tsunemi Kubodera captured the first still images of a live giant squid in its natural habitat—the deep sea. He outfitted a baited fishing line with a still camera programmed to take a picture every thirty seconds. He had been going out on Japanese fishing vessels for almost three years, to locations where he believed the squid might be found. The images he finally succeeded in capturing were a series of stills of a giant squid attacking bait on a hook at 2,952 feet.

When he released the images, the seismic public response helped motivate the Japan Broadcasting Corporation (NHK), with assistance from Discovery Channel, to finance the most ambitious attempt ever undertaken to film the giant squid in its natural habitat. And how I came to be part of that historic expedition was directly related to the success of the Eye-in-the-Sea.

After the triumph of the first Deep Scope mission, in 2004, I reapplied to the National Science Foundation, this time to produce a moored version of the Eye-in-the-Sea. Instead of having to deploy and recover the camera after only a day or two of recording, this was to be a semi-permanent installation in the Monterey Canyon. MBARI was taking the lead on developing what was essentially a power strip on the floor of the canyon, where scientists could plug in their gear and operate it remotely from shore via a thirty-two-mile cable.

To be able to observe unobtrusively for not just a day but *months* at a time would be a dream come true, so I was elated when, on this go-round, the reviewers gave the proposal an enthusiastic thumbs-up. The project manager for developing and building this new system was Lee Frey, the young engineer who had been so instrumental in getting the Eye-in-the-Sea operational at the outset. During those early days, he had to scrounge for parts and figure out inexpensive workarounds, but now that we had a budget of over half a million dollars, Lee was able to engineer a far more elegant system, with all the bells and whistles I had dreamed about.

In this new version, the camera was on a pan and tilt that could be controlled from shore, and there were three fold-down platforms that we still called CLAMs: one to hold a current meter and temperature probe; one to hold the e-jelly, a hydrophone, and bioluminescence sensors; and one to hold other experiments like bait that could be delivered by an ROV. It had both red and white lights that could also be controlled from shore and a set of parallel ranging lasers for gauging size.

The deployment of the camera in October 2009 went flawlessly, and when the ROV plugged it in for the first time, all of its systems came online without a glitch. After years of frustrations and setbacks, it felt like a miracle. But to Lee it wasn't a miracle at all. That would have required a deity; this just required the next best thing, a damn good engineer.

As soon as the system was activated, it automatically began recording, at which point our observations of life in the deep sea went from a paltry intermittent trickle to a gushing fire hose. I had anticipated this, and so a portion of the engineering budget was going to MBARI computer engineers to develop an automatic image recognition system that could identify the outlines of certain animals and flag the places on the recordings where there was movement, so we didn't have to spend hours scanning through video in which nothing was happening. The challenge, though, turned out to be that something was *always* happening.

There was a continuous flurry of marine snow and almost always a fish or crab or floppy sea pen in sight, and although they weren't extravagant in their behaviors—sometimes remaining in one place for hours at a stretch—that was valuable information for figuring out their energy needs. The computer engineers kept working on possible analysis techniques, but in the meantime I was trying to raise the funds I needed to bring the video stream to the web, making it the first deep-sea webcam. This was finally made possible through the graces of one incredibly generous and forward-thinking donor.

The animal behavior we were able to witness over the eight-month operation of the Moored Eye-in-the-Sea was the first long-term, unobtrusive observation of day-to-day life in the deep sea. We ran a variety of experiments. Those using the optical lure produced the most spectacular results, as we discovered just how attractive the e-jelly was to Humboldt squid. Again and again, when we ran the pinwheel display, we recorded Humboldts executing high-speed attacks. As soon as they realized there was no food associated with the light show, they would jet away, sometimes registering their displeasure in a cloud of ink.

There was one interesting exception to this routine, executed by what I dubbed the Einstein of Humboldt squid. For some reason, this squid recognized there was something off right from the get-go, and he approached with caution, hovering above the e-jelly, arms spread open, seemingly looking for the predator that was triggering this alarm call. Seeing none, he retreated, only to try again, and then again. After the third attempt he stayed away longer, presumably refiguring his approach, because on the fourth try he came in from the side instead of from above, but still finding nothing tasty, he gave up and jetted away.

Once I had the data streaming to the web, opportunities to observe such behavior proved to be highly addictive, not just to a handful of deep-sea biologists but to a wider and surprisingly diverse portion of the public. Word got out, and the interest was not just national but international; one father-and-son team in Italy watched the feed more than I did and had all sorts of interesting observations and questions. And after the camera was removed, I received many unexpected requests for its return, the most surprising one coming from a hospital that said it was the most watched internet feed on its cancer ward. Based on my own hospital experience, I can imagine why.

For a modern audience, often characterized as screen-addicted zombies who expect instant gratification, our deep-sea webcam would have epitomized eye-rolling boredom. Black-and-white video of a deep-sea flatfish sitting still on the bottom next to a floppy sea pen that was barely moving in the current, with occasional flocks of marine snow floating by, is not electrifying stuff. It's gripping theater only if you are able to step outside yourself and really think about what you're seeing. There is nothing like fighting for your own existence to make you appreciate the miracle of life. Once you do, then "watching and wondering" about life becomes a thoughtful, mind-absorbing exercise, heightened by the fact that at any moment you might witness something—an animal or behavior—that no one has ever seen before.

THIS MOORED VERSION of the Eye-in-the-Sea was a success. But it required an electrical power source cabled from shore, a hugely expensive and logistically challenging proposition. Funding was on the decline, and the opportunities for more such installations seemed bleak. The deployment of the earlier battery-powered version of the Eye-in-the-Sea was also becoming a challenge because of the expense of using either a submersible—assuming you could even find one—or an ROV to deploy and recover it. As a result, at the end of our 2007 Deep Scope mission, Justin Marshall, Sönke Johnsen, and I imagined a new kind of platform, which we called the Medusa.

The concept was to design something small enough that it could be dropped off the side of a ship; when it was time to recover it, an acoustic signal would release a sacrificial weight and it would pop to the surface, where it could be snagged and pulled aboard. Since it was based on the Eye-in-the-Sea, we agreed that Lee Frey was the perfect engineer to honcho the project and make it into the latest and greatest world-beating undersea camera; now we just needed to find some funding.*

Justin scored first, with money from the Australian Research Council as part of a push he had been leading for deep-sea exploration around Australia that he dubbed Deep Down Under. He raised enough to pay for Lee's engineering design and the production of two Medusa platforms. That meant I only needed to cover the costs of building a third, which was about $60,000. I applied to NSF with

---

*Cue the violins.

the goal of getting the Medusa built in time to use on a National Geographic–sponsored seamount mission to Costa Rica that I had been invited to join. Unfortunately, due to some delays in production, the Medusa wasn't ready in time.

As a result, its first deployment was on the site of the BP oil spill, on an expedition organized by Sylvia Earle* to survey the damage on the bottom, a few months after the gusher was capped. Bad weather plagued the expedition, and we managed only two deployments in shallow water, but they were critical to figuring out some problems and getting the Medusa fully operational. It was now all dressed up but with nowhere to go.

I BECAME A giant squid hunter as the direct result of a TED talk I gave in 2010. TED, which stands for "Technology, Entertainment and Design," is a remarkable organization with a simple mission: to spread ideas. In 2010, TED held its first conference at sea, aboard the 293-foot luxury expedition vessel *National Geographic Endeavour*.

The conference, billed as the Mission Blue Voyage, was part of Sylvia Earle's 2009 "TED Prize wish"—an award that the organization gives to "leaders with creative, bold wishes to spark global change." In order to explore ways to address some of the big challenges facing the ocean, Mission Blue brought together policy makers and influencers, leading scientists, innovators, activists, philanthropists, musicians, and artists for a nearly weeklong expedition to the Galápagos Islands in April 2010. It was an awe-

---

* Named one of *Time* magazine's first Heroes for the Planet in 1998, Sylvia Earle has spent her career working tirelessly for ocean conservation. I have many reasons to be grateful to her, but my fondest memory is of when I was diving Wasp. The light meter I was supposed to be testing was still on shore, and the sea state was deemed too wild to risk trying to deliver it by small boat to our research vessel. Sylvia, having overheard my ship-to-shore pleas to get it transported, broke in to volunteer as courier. I believe it was the first time we spoke.

inspiring event that interspersed TED talks with scuba diving, nature walks, snorkeling, music making, and boat trips. The power of being able to bring people to this special place and help forge an emotional connection to the sea was made manifest when participants committed $17 million to ocean conservation initiatives like the creation of protected marine areas, what Sylvia termed ocean "Hope Spots."

For my presentation, I spoke about the glories of bioluminescence, described the Eye-in-the-Sea, and showed some of the results we had been getting, using red light to be unobtrusive, in combination with the e-jelly as an optical lure. Another TED speaker was Mike deGruy,* one of the most exuberant advocates for the ocean it's ever been my pleasure to know. The TED talk Mike gave was a rhapsodic ode to the ocean in which he used no slides but painted vibrant detailed imagery with words.

After my talk, Mike was practically vibrating when he tracked me down to ask, "Do you think your red-lights-and-optical-lure approach might work for filming a giant squid?" I hadn't really thought about that, but okay, why not? "Sure," I said. "I think those enormous eyes suggest we should be paying more attention to its visual ecology. At the very least, we should *not* be using bright white lights that scare it away. And if it's an active predator, which I believe it is, then it might be attracted by an optical lure that imitates the bioluminescence it likely uses those eyes to see." I then proceeded to describe the Medusa and how we could deploy it either on the bottom or as a drifter.

Mike told me he was involved with a hush-hush project to try to film a giant squid for television and asked if I'd be willing to present my approach and findings at a meeting in Silver Spring, Maryland,

---

* Pronounced *degree*. I was familiar with Mike from his National Geographic series *Perfect Shark*, where he had done a segment on the bioluminescent cookie-cutter shark and showed an animation depicting my hypothesis of how the shark's "dog collar" might function as a lure.

in August. Mike's enthusiasm was indomitable and contagious, so, although the television connection gave me pause,* I agreed.

Dubbed "the Squid Summit," this was a gathering of television people and squid experts, including Tsunemi Kubodera, the Japanese scientist who captured the first still images of a giant squid in the deep sea; Clyde Roper, a world-renowned squid biologist from the Smithsonian who had been on several giant-squid hunts in the past; and Roger Hanlon, an expert in cephalopod behavior from the Marine Biological Laboratory at Woods Hole. The rest of those attending were television folks from NHK and Discovery. I got the impression that among the television contingent there was some skepticism about why I was there, and I suspected that Mike may have had to exert some significant pressure to get me in the door.

I made a science-oriented presentation with data to support the merit of my approach. I gave statistics on animals gathering around bait viewed under red light, as opposed to white light, and showed the importance of using red cutoff filters and intensified cameras. I displayed pictures of the Medusa and diagrams of its different possible deployment configurations, then I showed the e-jelly, along with the bioluminescent display it replicated, and talked about the importance of using an optical lure to attract active predators rather than just scavengers. I saved the video of squid attacking the e-jelly for the end. When they saw that, several of the television people leaned forward in their chairs. As I finished, I looked over at Mike, who was smiling. He knew they were sold.

The expedition was to take place off Japan the following summer (2011). I had a lot of misgivings about the whole operation, which would be leasing a privately owned vessel called the *Alucia*, along with its own submersibles, a two-person Deep Rover and a three-

---

*This was before Discovery Channel went over to the dark side with *Megalodon* and *Mermaids*. My reticence was based on my experience from the Cuba expedition.

person Triton. I had heard those were excellent subs, but knowing nothing about their crew was a source of consternation. There were only two reasons I hung in there: The expedition was going to afford me six weeks at sea, an unimaginable stretch of time under my current funding constraints, and Mike deGruy's enthusiasm was infectious. I figured that with him there to do most of the on-camera work, I could fade into the background and not be subsumed by the television production.

Given the series of events leading up to that expedition, it seemed predestined for tragedy and failure. On March 11, 2011, a massive earthquake measuring over 9.0 on the Richter scale, the fourth largest in recorded history, struck off the coast of Japan. Tsunami waves that reached heights of over 130 feet swept away whole communities and flooded the Fukushima Daiichi nuclear power plant, triggering three nuclear meltdowns. The tragedy was so massive and far-reaching that NHK was forced to postpone the expedition to the following summer, 2012. Then, on February 12, 2012, Mike deGruy was killed in a helicopter crash while filming a documentary in Australia.* His loss was devastating to everyone who knew him. At his memorial service, in Santa Barbara, Mike's brother Frank described his brother's life as being "one big human exclamation point!" A very bright light had been snuffed out.

Mike had brought me into this crazy project, and I had no appetite for carrying on without him. I wanted to bail on the whole thing—even more so when, around this same time, my technician, Brandy Nelson, who was the only person in my lab trained on the Medusa, informed me that she was pregnant and, since the expedition coincided with her due date, she wouldn't be able to accompany me. In the end, I went through with it, because I felt that if I didn't, it would be a betrayal of Mike. I reached out to Justin Marshall in

---

*The documentary, released in 2014, was *James Cameron's Deepsea Challenge 3D*. Also killed in the crash was Australian filmmaker Andrew Wight.

Australia to see if any of his people who had been trained on the Medusa could join us on the mission. Luckily, he had a Ph.D. student, Wen-Sung Chung, who was not only an expert with the system but also a squid enthusiast and eager to participate.

I DID NOT have a good feeling about what might be in store as David drove me to the West Palm Beach airport on June 19, 2012. I had had little communication with anyone involved with the expedition. The emphasis on secrecy and the loss of Mike as a go-between had left me very much in the dark. I found it tough to put on a happy face as I kissed David goodbye and hugged our golden doodle, Yankee, who had insisted on coming along for the ride to the airport. At the other end of this flight, there would be no one I knew. Even Clyde Roper, who I thought would never pass up an opportunity to go after a giant squid, had bowed out. The chief scientist on the mission was Tsunemi Kubodera, whom I had met only once, at the Squid Summit, and the only other scientist was Steve O'Shea, whom I had never met but knew by reputation to be somewhat eccentric.

Everything about the expedition was unconventional. Given my experiences with the Cuba mission, the emphasis on the needs of television production over the needs of scientific investigation was predictable, but some of the other anomalies turned out to be unexpected, in a good way.

For one thing, the *Alucia* was outrageous. A 182-foot luxury yacht with its own submersibles sounded like a James Bond fantasy when I first heard of it, but as soon as I saw her, she proved to be even more glamorous. Seen from the water, she looked impressive, with a large A-frame on the back for submersible launch and recovery* and a helicopter pad, which also served as the roof of her hangar, housing not just two, as I had originally been told, but three deep-

---

*In her previous incarnation, she had been the support ship for the French research submersible *Nautile*.

diving submersibles—more than any other ship in the world. In fact, the owner of the ship, hedge fund manager and philanthropist Ray Dalio, actually owned four submersibles: the Triton, two double Deep Rovers (one of which was to be used for filming establishing shots of the Triton; the other was in refit), and a double DeepWorker that was on board for this mission as emergency backup.

The inside of the ship was even more extraordinary. Accommodations on research vessels can be a bit spartan. I have had to share rooms so small they could be mistaken for closets, with bunks so closely stacked that when I turned on my side, my shoulder hit the bottom of the bunk above me, so when they told me that I would be sharing a room with two other women, Discovery Channel producer/director Leslie Schwerin, and Steve O'Shea's technician, Severine Hannam, I expected the usual tight quarters. But our room was positively palatial; instead of the usual portholes, it had panoramic picture windows looking directly onto the ocean; large, plush beds; scads of drawers and closet space with actual hangers;* a large desk; and an en suite bathroom that looked like something out of a Swedish sauna. Even more over-the-top was the news that the staff would be making up our bunks for us every day, doing our laundry, and serving us gourmet meals three times a day. Plus, it looked like I was going to get lots of time in the sub. Each day, the three-person Triton was to be launched for a seven-to-eight-hour dive, with a pilot, an NHK cameraman, and either Kubodera, O'Shea, or me as scientist observer. Six weeks of this was clearly not going to be a hardship.

On the other hand, I still had some safety concerns. The problem was that we were operating over very deep water. Submersibles deployed where the bottom exceeds their crush depths are supposed to be on a tether. My last experience with a tether was with Wasp, and it hadn't made me a fan, but at least with Wasp, that was its normal

---

*On research vessels that have locker space, you're lucky if you get any hangers, probably because they get scarfed up to jury-rig fixes on broken gear.

operational mode. For both the Deep Rover and the Triton, this was a whole new shtick and required devising a way to make it work.

The plan was to use a polypropylene line on a hand-cranked drum; the Triton's tether would be deployed from its usual tender vessel, the thirty-two-foot *Northwind*, while the Deep Rover's tether would have to be deployed from a Zodiac. The *Northwind* had the technology needed to track the Triton underwater, but the only way to track the Deep Rover was from the *Alucia,* which was going to require careful coordination between the ship's crew and the sub crew.

Of the forty-one people on board, eleven were Japanese, and most of them spoke little or no English. But since these were the folks footing most of the bill for this enormous endeavor, they were calling the shots. Discovery Channel was also involved and had three of its own people on board, but the level of the network's financial contribution was such that it quickly became apparent what the pecking order was going to be.

Added to all of these logistical concerns was the whole Hollywood Science aspect, which turned out to be worse than I had imagined. There were actually *two* documentaries being produced: the NHK version and the Discovery Channel version. I had been harboring a couple of illusions that were shattered on day one. The first was that Kubodera would be the focus of the NHK version and the second was that O'Shea, who clearly loved being in front of the cameras, would take on the brunt of the Discovery Channel version, filling the role that Mike deGruy would have taken. I was disabused of both of these notions the first morning on board, when I was asked to do an on-camera interview in the wet lab for the Discovery team.

It was then that I learned that they planned to frame their program as a competition between the three scientists, since each of us was focusing on a different approach. Kubodera was going to be using a large bait squid, like the one he had used to garner those first still images of the giant squid. The bait squid was to be attached via a line to the sub, and he planned to sit in the dark, using the red lights and the low-light camera to observe unobtrusively. O'Shea was planning to use ground-up squid squirted from syringes as a chemical lure, and I would be depending on the e-jelly as an optical lure—used on both the sub and the Medusa. All three of us vigorously objected to this story line, insisting that we would prefer to be viewed as working cooperatively, but our protestations fell on deaf ears.

At least the NHK team seemed to be focused on doing a more traditional natural history documentary, but even though it would be in Japanese, I wasn't off the hook because they wanted me to do on-camera interviews for them as well, which they would film in English and subtitle in Japanese. Although they weren't being quite as blatant as the Discovery team about treating this as a competition, that vibe was coming through, and I could tell that in both camps I was considered the long shot.

OUR DIVE SITE was off the Ogasawara Archipelago, a chain of subtropical islands six hundred miles south of Tokyo. It was in these waters that Kubodera had captured his still images and where he believed sperm whales came to feed on giant squid every year. The thing that had brought him to this location in the first place was local longline fishermen's reports of severed giant squid tentacles snagged on baited lures—in two cases, whole bodies—as well as descriptions of long squid tentacles dangling out of the jaws of breaching sperm whales. Shortly after we arrived, we spotted some of these toothed whales at the surface, easily identified by their lopsided bushy blow. It seemed we were in the right place.

Once we were onsite, we still had to test the tether mode, which was when I was fully expecting to see things go pear-shaped. To my great relief, I was wrong. It turned out there were several factors that made this a much smoother-running operation than I would have thought possible. The most important was the submersible team leader, Mark Taylor, a.k.a. Buck, a redheaded Brit with a mischievous grin. Buck's background with submersibles was impressive, starting with training as a diver and submarine operator for the Royal Navy; working on the rescue team for the Russian submarine *Kursk,* which sank in the year 2000; running submersible pilot training programs for the military and the private sector; and, judging by his sea stories, possibly racking up more experience with submersibles in hairy situations than anyone I've ever met. He definitely had the required eyes-in-the-back-of-the-head awareness, and his sense of humor went a long way toward alleviating tensions among the various factions on board. The sub crew were equally talented, and although they didn't all know one another initially, they soon were working together as though they had been doing so for years.

The day of the first science dive dawned clear and calm. Kubodera, as chief scientist, was first up. Talking with him and O'Shea over breakfast, I was surprised to discover that neither of them had been in a submersible before. O'Shea said he'd had opportunities but never wanted to go, his reason being "I'm a heavy smoker." Kubodera, who suggested we call him Ku (probably to escape various mangled pronunciations of his first and last names), seemed a bit nervous, but definitely game.

I love talking to people before and after they make their first dive, especially scientists who've spent their whole lives studying the ocean. It's fascinating to see how their perspective is altered by the experience. So I was looking forward to hearing Ku's impressions when he returned after that first seven-hour dive. Oddly, though, he didn't have much to say. He mentioned seeing a blue shark and that he was surprised by how much sunlight was still visible, even below two thousand feet, but when I asked about bioluminescence, he said there wasn't much. Seven hours of sitting in the dark while bobbing up and down on the end of a tether and he hadn't seen much bioluminescence? That seemed weird. Maybe he was just somebody who didn't show a lot of excitement, or maybe it was the language barrier. It was going to be a couple of days before I got to go myself. In the meantime, I was focused on prepping the Medusa for its first deployment.

Although I missed Brandy, it had quickly become apparent that Wen-Sung Chung knew his way around the Medusa. He had the camera system assembled and the float, weights, and lines ready to deploy with minimal assistance from me, but I was still so nervous anticipating our first deployment the day after Ku's dive that I awoke at three A.M. and never managed to go back to sleep as I ran through checklists over and over in my head. Although Wen-Sung had deployed the Medusa in drift mode, I never had, and I kept thinking of all the ways it could go wrong.

The theory was simple enough. Compared with the original Eye-in-the-Sea, the Medusa was compact, approximately two feet square and three feet tall, and light enough, at three hundred pounds, that its launch and recovery should be easy. We planned to just launch it off the fantail using the A-frame. Once it was released from the Sea Catch,* it would drop toward the depths, paying out the 2,400 feet of line that attached it to a float with a satellite tracking beacon at the surface.

As soon as the Triton was in the water for its second dive,† we launched the Medusa. Once it was released, Wen-Sung and I piled into the Zodiac to follow the float so we could use the acoustic transducer to watch it drop through the water. This was mostly out of an abundance of caution, so that if it came off the line and was dropping toward its crush depth, we could send the signal that would cause it to drop its sacrificial weight and come back to the surface. As it dropped slowly through the water, we kept pinging it to check its depth, and when it approached 2,400 feet I held my breath.

---

* A quick-toggle shackle-release mechanism.

† For this dive the scientist's seat was filled by a Japanese scientist who was an expert on the acoustic imaging system that was being tested as an alternative method of observing without white light.

When the next ping gave a reading of 2,460 feet, I freaked. *Seriously? Lost on the first deployment?* But when I looked at Wen-Sung, he was smiling; he pointed out that we had drifted away from the float, so the distance we were measuring was not the depth (the distance between the Medusa and its surface float) but rather the hypotenuse of the triangle formed by the Medusa, its float, and our Zodiac. Obviously he had seen this before. We took a few more readings to make sure it was stable and then powered down the transducer.

That imagined near miss stuck with me, though, and once we were back on the ship, I found I was suffering from severe separation anxiety, envisioning the whole slew of ways I might never see the Medusa again. Buck agreed with my desire to have a plan B, so after the Triton was back on deck he helped round up some spare gear and then accompanied us in the Zodiac back out to the float, where we attached a second float, a VHF tracking beacon as a backup for the satellite beacon, and a strobe, all of which made it possible for me to sleep that night. I dreamed about the Medusa drifting through the dark depths with its far-red lights and intensified camera probing the darkness.

The next morning it was my turn to dive in the Triton. The plan was simple: Drop down below the edge of darkness and then sit with the e-jelly on the end of a long pole in front of the sub and watch for attacks by way of the intensified camera and red-light illumination. As I nestled into the cushy starboard seat, my excitement was acute. It wasn't just the possibility of seeing a giant squid but the unprecedented opportunity for long-term unobtrusive observations in the midwater that had me amped up.

Even if we didn't find a giant squid, I was absolutely sure we would see something amazing. Which is why, as the hours ticked by and we kept seeing a whole lot of nothing, I became more and more dismayed. The waters were empty—some of the most barren I've ever been in. They were crystal clear, with hardly any marine snow and even less bioluminescence. It made no sense. Where was the food web needed to support giant squid and sperm whales?

Maybe things would look different on the Medusa footage. We were scheduled to pick it up right after my dive, but it took a while to recover, because it had drifted so far north from where we were diving. Once we retrieved it and got it back on deck, I was enormously relieved to see it had taken data, but when we did a quick scan through the video, there was almost nothing—a siphonophore and a couple of shrimp. The next two submersible dives, first by O'Shea and then Ku again, were equally discouraging. There was palpable gloom surrounding the Japanese organizational meetings held in the lounge every evening. NHK had made an immense investment in this endeavor, and there were many careers riding on the outcome.

My second dive in the Triton was on July 3. In the interim we had deployed and recovered the Medusa a second time, now with the e-jelly in place, which had been left off the first deployment because it still needed to go through a pressure certification. I had taken a quick look at the video the night before and seen that the water looked as empty as on the first deployment. Wen-Sung planned to review the footage in detail while I was in the sub. At least during this dive I saw some bioluminescence, including some flashback, but not much else. We were back on deck by 3:30 P.M. I did an interview on the fantail for NHK, talking about the bioluminescence I had seen, and then I retired to the mess to rehydrate with a cup of tea.

Steve O'Shea tracked me down there to say, "Wen-Sung found something on the video he wants you to see." He didn't seem that excited, so when I sat down next to Wen-Sung I was expecting a jelly or shrimp that needed identifying. The Discovery team was filming us, which was not unusual. Wen-Sung pulled up a couple of pretty unexciting sequences and then a long stretch of empty water, and I was thinking *a whole lotta nothing* when suddenly, three enormous squid arms swept into view from the right-hand side of the monitor. My heart did a somersault as they cascaded across the screen between the e-jelly and the camera lens: thin tips at first, widening into thick, muscular arms that arched and flexed, simultaneously powerful and supple. They weren't round in cross section but triangular, with two rows of protruding suckers running along the base of the triangle. Under the red-light illumination with the black-and-white camera, they looked as white as Moby Dick.

I shouted, "Oh my God!" and looked around for Ku to confirm this was the real thing. The holy grail! The first video of a giant squid filmed in its natural habitat? Ku, O'Shea, and Wen-Sung were grinning like jack-o'-lanterns and we all basically lost our minds, jumping around and yelling and hugging.

There were actually three separate sequences of the squid entering the frame. The first two were only four minutes apart, the third more than an hour later. In two of the recordings, we saw only its arms up close, but in the third the whole animal was visible, hanging in the distance—a dim grayish outline—arms and tentacles splayed out like a half-open umbrella. Years and years of unsuccessful hunts for giant squid, and the first time the Medusa is deployed with the optical lure it scores *three* sightings! It was the sweetest victory I had ever known. To be able to explore in this new way—attracting rather than repelling such a famous, long-sought quarry—and have such an unqualified success was beyond my wildest imaginings.

After seeing how much the squid overfilled the screen, I decided we needed to try and get the e-jelly farther away from the camera, so Wen-Sung and I replaced the two-foot-long aluminum strut holding the e-jelly with a three-footer. On the next Medusa deployment, we scored yet another giant squid sighting. This one was three miles away from the first, and five days later. We didn't know whether it was the same squid or a different one.

The mood on board was teetering on giddy. This being a television production, there was now a lot of emphasis on getting B-roll of the Medusa during launch and recovery. There was also greatly increased hope that we might actually still film a giant squid from the submersible, using the same approach as with the Medusa—red illumination and an optical lure—but with the added benefit of high-resolution color cameras.

It happened only a few days later, during Ku's dive. As bait, he was using a three-foot diamondback squid that local longline fishermen had caught. To optimize its presentation, it was attached to the sub with almost fifteen feet of monofilament, and its body cavity was stuffed with a few blocks of syntactic foam, making it only slightly negatively buoyant so that it would sink slowly through the water. It was also now rigged with an optical lure, in the form of a deep-sea squid jig that flashed blue, green, and red. The other passenger in the sub was NHK cameraman Tatsuhiko Sugita, known as "Magic Man" because of his talent for doing magic tricks, with which he had impressed us all at a couple of parties. The pilot, Jim Harris, had just come on board the day before to replace another pilot who was leaving. It was his first dive for this expedition.

The day started off uneventfully. Launch and recovery of the sub had become so routine that there were few people out on deck to see Ku, Jim, and Magic Man off for their early-morning launch. The excitement started after lunch, when I heard that Jim had notified control that they had filmed a giant squid. I rushed to the control room to try to get more details, and as soon as I entered, I could feel the tension. Floodlights were on and all cameras were rolling, apparently filming the reaction shots of those in the control room—but reactions to what?

At the moment, nobody seemed to be saying anything and I still wasn't clear on why they were continuing to film. When they said they had filmed the squid, I assumed the encounter was just a few seconds, but apparently it was still going on. When I asked, "For how long?" the response, "Fifteen minutes so far," left me incredulous—it had to be a miscommunication. But then I heard Jim's voice over the comms, saying "Still filming." In total, they filmed the giant for twenty-three minutes! There was no direct video feed from the submersible to the ship, so we had no way of seeing. I was mystified about what could have been going on for that length of time.

Everyone turned out on deck to greet them for their triumphant return. With all the excitement and talking at once, it took a while to extract the salient details, but the gist of it was that they went down with only the red lights on and with Jim carefully matching the drop rate of the sub to that of the bait squid—no thrusters, just ballast control. As they sank past the edge of darkness, the red light was too dim to see the bait squid except on the intensified camera, so Jim said he had used the small flashing jig light to judge the location of the bait. They were around two thousand feet deep when the giant attacked.

We had to wait till that evening in the lounge to see what they were trying to describe with words. Under the red-light illumination, we could see the squid attack, spreading its eight arms wide to engulf the bait at the head end, farthest from the sub. While the camera was picking this up, Ku was in the submersible, peering into the darkness, but could see nothing. Excited and desperate to get a better look at what was going on, he turned on his white flashlight. When that didn't scare the squid away, he decided to risk turning on the submersible's white lights. Watching the screen, I felt like I was right there with him. When the lights came on and the screen whited out, I was holding my breath. Then, as the camera's automatic gain control kicked in, the squid came into high-resolution focus. It was *magnificent*. And utterly at odds with what I was expecting.

The most striking first impression was its metallic coloration, which seemed to shift between brushed bronze and polished aluminum. That was a complete surprise, since the dead and dying specimens seen in the past were all red. And the coloration kept shifting—from predominantly bronze to mostly silver. The arms had a very distinctive triangular shape and undulated in the current. Their color was grayish white, with randomly spaced bronze horizontal stripes almost like a bar code.

The eye staring at us was enormous and alien. It was almond-shaped, with a huge area of white surrounding what looked like a narrow iris and a very large black pupil. It was an eye you could get lost in. Initially the eyeball was rotated in its socket such that it seemed to be looking away from the sub, possibly to avoid the bright lights, but later it looked directly at the camera. When it did, Ku said, "It looks rather lonely." But my assessment was that it was hungry, which I think is why it wasn't scared off when the white lights came on. Animals have hierarchies of behaviors, and in the case of this squid, I believe once it started feeding, the biological imperative to eat overpowered its instinct to flee.

It was vertical in the water, with its head up, fins flapping slowly while it held the bait squid inside the base of its thick, muscular arms, the ends of which were bent up and to the side as it dropped through the water. We estimated that the length from the tip of the tail to the ends of the arms was about ten feet. That meant that if its tentacles were fully extended, it would be more than twice that length, making it as tall as a two-story house.

We watched the screen with rapt attention. Still, the squid continued munching on its feast while Jim adjusted the drop rate on the sub to keep falling in tandem with the squid as Magic Man filmed it with the high-resolution cameras, getting both long shots and close-ups. He filmed until the end of the tether, at three thousand feet, stopped their descent, at which point the giant apparently sensed a change that caused it to drop what was left of its feast and jet away into the darkness.

For most everyone on board, that footage was the climax of the expedition and obviously the capstone for the documentaries. But for me, there were two more high points still to come. The first was another recording of a giant squid made during the fifth Medusa deployment. This sequence was high drama—a full-on attack where the whole giant squid was visible as it swooped in with arms and tentacles held together like a spearpoint.* It seemed to be going for the e-jelly, but then at the last instant it arced up and over it, spreading its arms wide to embrace the Medusa, thereby exposing its mouth directly to the camera lens. It was exactly the behavior you would expect to see from a secondary predator responding to a burglar alarm: The squid initially homed in on the light display, but at the last moment diverted its attack to the big thing next to it, which it presumably viewed as the primary predator that had elicited the alarm.

The second high point occurred about a week before the end of the expedition. I was still trying to get a handle on how these waters could sustain giant squid and sperm whales and yet appear so devoid of any other life. I had done one bottom dive and established a general paucity of life down there as well, so upwelling of nutrient-rich water wasn't the cause. My best guess was that there must be plankton-rich eddies spinning off from the Kuroshio Current, which flows along Japan's southeast coast. Also known as the Japan Current, it is similar to the Gulf Stream in the North Atlantic, an enormous river in the ocean that transports tropical water northward. Swirling off from the eastern edge of these currents are isolated rings or eddies that can be more than a hundred miles in diameter and constitute unique ecosystems. These eddies provide rich hunting grounds for predators and seemed to me to be the most likely food source.

I had assumed we would have access to satellite data on the expedition, but that turned out not to be the case, so there was no direct way to pinpoint an eddy in real time. Nonetheless, I had been campaigning for a dive farther north, which is where the Medusa had drifted when it recorded those giant squid sequences and where we would be most likely to encounter an eddy. I also wanted to dive at

* In the documentary they showed this out of order, before the recording from the submersible.

night, with the hope that we could record some bioluminescence. When I finally got that dive, I was rewarded almost immediately with intense bioluminescence at 550 feet, produced by a thick layer of krill. There was also impressive flashback luminescence in a broad layer between 1,200 and 2,100 feet. Here, at last, were the food-rich waters needed to sustain these giants.

Even better, just below 1,000 feet, the largest squid I've ever personally seen from a sub sped by so close that I felt like I could have reached out and touched it. At first, given its size, I thought it was a giant squid, but then I realized it was something equally thrilling for me—the octopus squid (*Taningia danae*). They are so named because, although juveniles have the usual two tentacles plus eight arms, mature specimens, which can exceed seven feet from tail to arm tips, generally lack the tentacles. What they have instead are two of the largest and brightest bioluminescent light organs found in any animal. Located at the ends of two of its arms, they are the size and color of lemons, although the light they produce is blue.

For this dive, we were replicating the protocol that Ku had used, but this time with a bait squid and the e-jelly. The big payoff to that approach came an hour and a half after that first sighting when we saw it again, or another just like it, at 1,338 feet. This time, the octopus squid attempted to grab the bait squid and tugged so hard on it that we felt the jolt in the sub. Contact! An exhilarating, dramatic end to my final science dive of the expedition.

On our last day on our dive site, we held a memorial service for Mike deGruy, with those who knew him gathering on the fantail at sunset. He should have been there for this enormous victory. People shared stories of Mike and talked about the phenomenal positive energy he brought to the world. I intended to speak, but when my turn came, the words caught in my throat. I felt bad about that in the moment, but later found a way to make up for it when I gave a TED talk, "How We Found the Giant Squid," which I dedicated to Mike. It's been viewed over five million times.

* * *

GIANT SQUID: THE MONSTER IS REAL was the title that Discovery Channel proposed for its documentary, which was scheduled to air six months after our return. Ku, O'Shea, and I objected vehemently to this characterization, on the grounds that of *course* it's real—scientists have been studying dead specimens since the mid-1800s. And, just as important, it's not a monster!

Like so many of history's terrors, once confronted, this fabled behemoth turned out to be a rather shy giant, unaware of its fame and scurrilous reputation, hiding in the dark depths, instinctively avoiding the bright lights of our exploration platforms possibly in the same way it escapes from the broad-field illumination stimulated by attacking toothed whales.

Further evidence of its mischaracterization was visible when we recovered the sub after Ku's dive. The bait squid was still attached. Incredibly, given the twenty-three minutes the giant squid had been feeding on the diamondback carcass, there was still a lot of it left. Looking at the marks on the mantle, they seemed more like dainty nips than bites, completely at odds with the flesh-ripping horror so often depicted.

Clyde Roper, the famed giant squid hunter, who was brought in to see and comment on our footage for the documentary, was incredibly gracious about our success. He also helped our case by adding his objections to Discovery's choice of title. Finally, a "compromise" was brokered. They capitulated by changing the title from *Giant Squid: The Monster Is Real* to *Monster Squid: The Giant Is Real*. Their rationale was that "monster" in this context was being used as an adjective referring to its size rather than its morals.*

NHK and Discovery managed to keep our achievement a secret

---

*It was remarkably reminiscent of the Cuba expedition, when Discovery magically altered the meaning of "forbidden" by changing the title from *Cuba: Forbidden Waters* to *Cuba: Forbidden Depths*.

right up until their documentaries aired in 2013. Promotions started just before the air date, and there was major public interest, with numerous giant squid parties and celebrations across the country for the premiere. I know because people sent me pictures of their giant squid piñatas, cakes, art, and tattoos inked in honor of the unveiling. I was, frankly, stunned by the intensity of interest from the public at large.

Nonetheless, I thought that once the documentary aired, the fervor would die down. But that wasn't the case, as I discovered in 2019, when I took the Medusa on a very different voyage, a NOAA-funded science mission in the Gulf of Mexico that we called Journey into Midnight.

SÖNKE, TAMMY, AND I were still rotating chief scientist duties for these Deep Scope visual ecology missions, and it was Sönke's turn to take the lead. We were going out on the 135-foot R/V *Point Sur*. There was no submersible and there were no television people. Our exploration tools were the remotely operated vehicle *Global Explorer*, a midwater Tucker trawl, and the Medusa, rigged for drift mode.* We had assembled a fantastic team of folks, with the goal of exploring the open-water environment below 3,300 feet, the least explored zone of the ocean. To assist me with the Medusa for this mission, Nathan Robinson† joined our team.

We launched the Medusa on Monday, June 17, 2019, for its fifth and shallowest deployment yet on this cruise. The other four had all

---

*This was its first deployment in the midwater since the Japan trip.

† If you ever saw the video that launched all the say-no-to-plastic-straws campaigns, that was Nathan Jack Robinson pulling the straw out of the turtle's nose. At the time of our mission, Nathan was the director of the Cape Eleuthera Institute, in the Bahamas. When the Medusa isn't needed for an expedition, I loan it to CEI for its deep-sea shark research and to help foster a new generation of explorers—high school students from the CEI-affiliated Island School, which provides these teens the opportunity to participate in real science.

been below 3,300 feet. At 2,500 feet, this one was close to the same depth we had been at for the giant squid recordings off Japan (2,400 feet). We recovered the Medusa late on Tuesday, our teammates taking breaks from their own research to help haul in the line— participating in a friendly competition to see who could coil it into boxes the fastest. After dinner, Nathan began downloading the video.

Late into the night on Tuesday and through the following morning, Nathan and I traded off reviewing the recordings on his laptop. We had gone through more than twenty hours of footage when, a couple of hours into his shift, Nathan came to find me. Saying nothing, he signaled me to follow him, and I knew by the expression on his face that he had discovered something. I shadowed him into the lab and watched over his shoulder as he hit play.

At first there was nothing except marine snow drifting by horizontally, making it apparent that the Medusa was being pulled by the current. Then, from the left-hand side of the screen, a squid appeared. It was jetting along horizontally, arms first, keeping pace with the e-jelly. As wave action at the surface was transmitted down the tether, the Medusa and its attached e-jelly oscillated gently up and down, and in lockstep, the squid undulated up and down, too. It appeared to be visually tracking the e-jelly!

The attack, when it came, seemed almost casual. The squid bent its arms, held together in a spearpoint, toward the e-jelly, and when it made contact, the arms splayed apart, going every which way. The suckers of one glommed onto the side of the e-jelly while others stroked it and the bait bag attached next to it; tasting and testing and finding our offering unsatisfactory, it then let go and swam away.

Nathan and I started hollering like maniacs, bringing scientists and crew running from all over the ship. Again and again we played the sequence, trying to estimate the size and sort out other key taxonomic features. This would be only the second expedition ever to capture video of a giant squid in the deep sea. If the Medusa had really done it again, and in our own backyard, only a hundred miles southeast of New Orleans, that was a huge deal and a clear testament to the first time not being just luck. But before we made any announcements, we needed to be sure. We wanted to send the video clip to Mike Vecchione, the squid expert at the Smithsonian, for confirmation. To our extreme frustration, though, we couldn't, because we were in a nasty squall and the internet was down.

We were discussing the best way to estimate the squid's size when the lightning struck. It was an earsplitting crack that sent us all scrambling out on deck. The ship's long-range antenna had been hit and blown to smithereens. The destructive force of a direct lightning strike is truly awesome. Electronic devices—computers in particular—are especially vulnerable, which is why, as soon as we realized we'd been hit, Nathan and I raced back to the lab to check on his laptop. One of the ship's computers was fried, but Nathan's, miraculously, was fine. Then, just as we were congratulating ourselves on having dodged a bullet, the ship's captain came down from the bridge to warn us that there was a waterspout forming off our port bow—a big one. It was as if we had summoned Poseidon's wrath for trying to reveal his leviathan to the world.

Thankfully, the spout gave us a miss, and the seas began to calm. We kept checking the internet connection and, in the meantime, went back to trying to estimate the length of the squid. Sönke described it as being like trying to measure an elastic band as it's fired directly at you. Based on the known size of the e-jelly, we came up with a conservative estimate of maybe ten feet, but later, when I could do more careful measurements and calculations, I concluded that it may have been more than twice that.*

Once the internet was working, we sent the video to Mike and then waited anxiously for his pronouncement. When he wrote back with a thumbs-up, we were ready to share the news. Sönke, Nathan,

---

*Although no flagons were involved, skepticism *is* warranted.

and I gave an interview to *The New York Times* over the *Point Sur*'s satellite link. At the same time, we posted the story and video on NOAA's Journey into Midnight website and immediately it went viral. News agencies around the world carried the story and we were bombarded with more interview requests.

Based on the amount of coverage our sighting received, this was a far greater public response than had occurred after the Japan expedition. Because there was no documentary this time around, we didn't have any of the cool B-roll footage to flesh out the story—like our on-camera reaction to the first sighting—but being able to share the discovery in near real time and not having to restrict the availability of the squid footage more than made up for that. It also gave us the freedom and opportunity to bring up important ocean issues in a context likely to reach a wider audience than the usual doom-and-gloom diatribes.

In our blog on the NOAA website, Sönke and I pointed out that our sighting had occurred at the edge of the Gulf of Mexico oil field, which produces nearly two million barrels of oil per day. In fact, we were so close to the Appomattox deepwater oil rig, one of the largest and deepest drilling platforms on the planet, that we could see them burning off methane every night at sunset. The human imprint on the planet, resulting from our continued dependence on fossil fuels, extends even into the lair of the legendary Kraken!

In interviews, I tried to emphasize how little we have actually explored of our own planet. Based on the large quantities of giant squid beaks found in sperm whale stomachs and given how readily we filmed them with the Medusa, it appears that giant squid aren't rare; they're just shy. We only knew about their existence because they happen to float when they die.* How many other amazing creatures

---

*This is because they have ammonium in their body tissues, which is unusual except in a few deepwater squid, and why you don't have to worry about seeing "all-organic, free-range Kraken" showing up on your menu now that we know how to find them.

are down there that we don't know about because what little exploring we have done we've done wrong?

And how many other terrors have been simply misidentified? For
centuries, the Kraken was reviled as a terrifying monster, but on
closer inspection it proved to be not monstrous but magnificent. For
most of human history we have viewed nature as a monster to be
battled and beaten into submission. In *Moby-Dick,* Ahab refused to
give in to nature's dominance, symbolized by the great white sperm
whale, which he saw as evil. For contrast, Melville offers up the viewpoint of another whaling captain, one who lost an arm to Moby Dick
just as Ahab lost a leg, but nonetheless viewed the creature as lacking malice, and counseled Ahab to leave the whale alone. In the end,
Ahab's egocentric obsession was his downfall. The whale destroys
Ahab, his ship, and all but one of his crew. As our numbers and our
destructive power on the planet grow, I fear that if we persist in seeing nature as a monster to be subdued, we risk Ahab's fate.

Chapter 14

TALKING TO CANNIBALS

The first Humboldt squid emerged from the gloom at 1,090 feet. It was arcing diagonally downward from left to right, arms first; large triangular fins curled over its back, barely moving. The appearance of speed without effort stemmed from its highly efficient jet propulsion system. I got only a tantalizing glimpse before the squid plunged into the blackness below the Deep Rover's red lights, but the impression was of mass and strength.

The next squid appeared at 1,475 feet. This one came out of the dark above us, targeting the electronic jellyfish affixed to a long pole in front of the sub. The Humboldt knifed in with all eight arms held together in a sharp point and then at the last second curled its arms back and spread them apart to engulf the e-jelly. Big and powerfully built, its body almost filled the field of view vertically. Discovering a piece of machinery instead of prey, it immediately reversed course by flapping those enormous fins while rotating the direction of its funnel. "Now we're talkin'!" I whooped.

This was really going to work! The idea was to use the e-jelly to draw in Humboldt squid so they could be filmed for the latest version of the BBC nature documentary series scheduled to air at the end of 2017 as *Blue Planet II.** In terms of TV appeal, these squid have it all; they are big, aggressive, Shaq-sized predators with an extensive behavioral repertoire, including a phenomenal capacity for visual communication. Not only can they use their whole body like a billboard, strobing colors and patterns in the blink of an eye, from red to white and back again, by contracting muscles around tiny pigment sacs called chromatophores, but they also have photophores that emit blue light, suggesting a capacity for an entirely different kind of light show.

We were hunting off the coast of Chile in a stretch of ocean that is part of the biggest invertebrate fishery in the world. Humboldt squid are the basis of this reputation. There are so many squid in these waters that they are harvested at the rate of hundreds of thousands of tons per year, putting food on dinner tables around the world. Your restaurant order of calamari with parsley and garlic may well be a portion of one of these seven-foot-long predators, which feed in swarms to the point of frenzy and are known to be so aggressive they'll attack and eat each other in the process.

To film the squid, we were using the same ship, the *Alucia*, and the same two submersibles from our Japan expedition: the three-person Triton and the two-person Deep Rover. In this case, the two observers in the Triton were BBC producer Orla Doherty and cameraman Hugh Miller. The Triton was outfitted with not one but two high-resolution cameras, one with ultra-low-light sensitivity, both mounted on the front of the sub. Deep Rover carried pilot Toby Mitchell and me, along with powerful external lights, both red and white, for illuminating whatever the Triton's cameras were filming. This entailed coordinating our filming efforts such that Toby would line up the Deep Rover to backlight or side-light whatever Orla and

---

* Described by *The Atlantic* (January 16, 2018) as "the greatest nature series of all time." (Also credited with stimulating a swell in applications for marine biology studies.)

Hugh were trying to film, in order to ensure maximum contrast and minimum backscatter.

My official title for this expedition, according to the ship's roster, was chief scientist. Orla simply referred to me as "the squid whisperer." I felt honored by the latter title, and undeserving of the former: Having been chief scientist on many submersible expeditions, I am all too familiar with the stress and exhaustion that normally accompany that moniker. Usually it involves writing grant proposals, acquiring the permits for operating in foreign waters, assembling the science team, coordinating travel and accommodations for said team, managing interactions between the science team and the ship and submersible crews, and trying to get everyone to play nice. On this expedition, though, I had none of those responsibilities. BBC and its collaborators, Alucia Productions, had assumed those duties, which meant I could just focus on the fun stuff: exploring.

As for "squid whisperer," after so many years of trying to decipher the language of light—from the early failures in Wasp and then Deep Rover, through the triumphs with first the Eye-in-the-Sea and then the Medusa, to that thrilling moment at the close of the Japan expedition when I not only *saw* but *felt* that large octopus squid attack the e-jelly—this had the ring of welcome validation.* On this expedition, I would not only have the opportunity to repeat that in-person thrill, possibly many times over, but I was also hoping to observe a form of bioluminescence communication never witnessed before.

My scheme for enticing the Humboldt squid into the range of our cameras was the same one we used off Japan: dangling the electronic jellyfish out in front of the sub while we and our cameras observed unobtrusively, using red light. I hadn't brought the Medusa for this mission, so everything depended on what we were able to see and record from the subs.

---

*Never mind that the only word I know how to whisper in Humboldt Squid–ese is "Lunch!"

Our first dive site was thirteen miles offshore, where the bottom depth was about three thousand feet. We had launched at four P.M., with the goal of exploring the water column both before and after sunset. My prediction was that we would find the squid at that sweet spot where sunlight relinquishes its hold, at which point I imagined the squid would be poised to begin their evening vertical migration to the surface.

Shortly after we began our descent, I realized I needed to revise my expectations as to depth. These were very different waters than I was used to. Here, the top two hundred feet contained a blanket of plankton so concentrated and thick that it swallowed up the sunlight. By the time we reached three hundred feet, the light was nearly gone. In the crystal-clear waters of the Bahamas, where I have made hundreds of submersible dives, we never saw daytime light levels this dim until we were below two thousand feet.

The richness of life in these surface waters is what makes this fishery so productive. Like the Gulf Stream and the Kuroshio Current, the Humboldt Current, from which the ravenous squid gets its name, is a major part of the ocean's circulation pattern, flowing north along the west coast of South America. The current creates upwellings of nutrients, which are the building blocks of life and the foundation for the exceptional biomass found here.

I knew this in theory, but seeing it in practice was still a surprise. The water was a dark teal green at the surface, rich with phytoplankton, and at 150 feet it was jelly soup—a gumbo of gelatinous life-forms feeding off the phytoplankton and one another. But as we descended farther, that profusion of life began to thin out. By 700 feet, any signs of macroscopic life disappeared and the water turned milky. This was the center of the oxygen minimum zone (OMZ)—the yin to the surface layer's yang.

Everything that lives must eventually die. As the phytoplankton and zooplankton at the surface expire and sink toward the bottom, they are consumed and decomposed by microbes. This process uses up large quantities of oxygen. In some regions, usually along the western coasts of continents, a lack of mixing can lead to a distinct layer of water with insufficient oxygen concentrations for most of the ocean's larger life-forms, a wet desert that fluctuates in size, depth, and range.

We were far below sunlight's reach, just coming out of the bottom of the oxygen minimum zone at about 1,090 feet, when we saw that first Humboldt. As we continued downward, we passed through a scattering of lanternfish, a common prey item for Humboldt squid. Lanternfish, so named because their bodies are studded with light organs, are found anywhere the water is deep and salty, which is more or less most of the world's oceans. Here they seemed to be concentrated in a layer between 1,800 and 2,100 feet. As we sank through this layer, we saw seven more Humboldts, with the last two at around 2,070 feet, one of which attacked the e-jelly. After that, the animal life thinned out until we approached the bottom, at 2,990 feet.

It had taken us more than an hour to journey a distance of less than twelve city blocks, but we had gone through the looking glass to another world. Here we found an abundance of a very different kind of life, in the form of the largest number I'd ever seen of a creature first described by ROV pilots as "headless chicken fish." In fact, they're *Enypniastes,* a type of pelagic sea cucumber, and I actually think they're beautiful. Granted, if you squint, the animal's body does bear a striking resemblance to a headless, wingless plucked chicken, but it swims by means of a large, webbed veil at its front end that pulls it through the water with balletic grace. The veil first points forward and then spreads out like a fan and curls backward, pushing water out between the veil and its body.

Their color is also very un-chickenlike. They are semitransparent, with a pinkish or reddish hue when seen under white light. Even more fantastic is the sticky blue bioluminescent fairy dust that rubs off their bodies upon contact, which serves as a defense against predators in the same way that exploding paint packets thwart bank robbers. Any predator foolish enough to try to chow down on one of these "chicken fish" would find itself with the equivalent of a glowing bull's-eye painted on it, making it an easy target for its predators.

I had often seen *Enypniastes* as solitary swimmers or in small groupings of two or three at a time, but here there were hundreds. It looked like a hot-air balloon festival, all these chicken fish drifting or swimming at different heights while others grazed on the bottom, rolling out lacy-edged feeding tentacles that gathered sediment into their mouths. They are the vacuum cleaners of the seafloor, and their abundance made perfect sense, given the dense rain of detritus from above.

I wanted to stay and enjoy the ballet, but we had gotten the disappointing word from mission control that the weather was picking up and we were being called back to the surface early. It was now approaching sunset, and as we ascended, we found that the lanternfish had started to migrate up. Clearly, in these inky dark waters, their diurnal journey wasn't triggered by light, so I wondered if their internal clocks were providing the migration cue. From the fish's minimum depth of 1,800 feet on our descent, they now extended up to 1,000 feet, and it was in their upper range, between 1,000 and 1,300 feet, just below the OMZ, that we again found the Humboldt squid. They were now hunting in earnest.

It was a different kind of attack from the one we had seen earlier on the e-jelly. It started out the same, the squid homing in on the fish, arms held together in a tight point, but then the eight arms splayed apart and two elastic tentacles shot out, grabbing the fish, which flashed a bioluminescent scream for help. At the same time, some of the squid produced a stroboscopic visual effect, changing their entire body color from red to white, shifting back and forth two to four times per second—a hugely intimidating display in such a large, aggressive predator. Intriguingly, it never seemed to happen when we saw just one squid alone. Strobing occurred only when there was more than one Humboldt in sight. This was communication—the transfer of information—and it was about as subtle as a jackhammer. But what were they saying?

When cannibals chow down, it's pretty important that there be no miscommunication. If two cannibals try to attack the same prey at the same time, the loser might just decide to attack and eat the winner. In squid-on-squid aggression, size differences play a big part in determining who is the diner and who is the dinner. Whole-body strobing is one way to convey intention—*I'm about to attack this fish. It's mine. Stay away!*—as well as vital size and strength data: *Back off, buddy! I'm bigger than you.*

We saw more than thirty squid on that first dive, which for me was nirvana. But for Orla and Hugh, in the Triton, it was frustrating to be close enough to see the squid, yet so far away that we couldn't get the kinds of shots they needed. Still, at least we knew the squid were here. Hopefully, all we needed to do was be patient and keep trying.

However, with subsequent dives we kept seeing squid, and they were consistently being drawn to the e-jelly, but we weren't getting the shots Orla needed. The problem was by design, sadly. Because the burglar alarm is a last-ditch scream for help, secondary predators must respond quickly, before the primary predator extinguishes its landing beacon by consuming the prey and departing. This meant that the Humboldts homed in on the e-jelly at high speed and, when they found nothing to eat, zoomed out again just as fast.

In an attempt to get the squid to stay longer, we tried attaching a bait squid next to the e-jelly. This produced what I considered cinematic gold. A hefty seven-foot Humboldt came in on the e-jelly and grabbed on to the bait squid, attempting to wrench it free, while its whole body pulsated back and forth between red and white. The sub shook with the force of the squid's yanks, and the brute did not give up easily, which meant we got a lot of footage. This seemed ideal to me, but Orla feared she couldn't use it because it would not be deemed natural behavior by the *Blue Planet* team.

On another occasion, during one of the few dives when I wasn't in the sub,* they actually managed to film an incidence of cannibalism—the first time this behavior was ever recorded! A large squid grabbed a smaller one and then spewed out a smoke screen of black ink to try to hide its prize, but to no avail, upon which an even larger squid swooped in to grab it and, after a brief tug-of-war, stole it away. This whole scene was filmed in close-up, which was phenomenal but very brief. Much more was needed to tell the story.

And suddenly it was our last dive. We had a long list of things we needed to accomplish. My number-one priority was filming bioluminescence. I had managed to convince Orla that an important part of the Humboldt squid story was figuring out what part bioluminescence plays in its visual communications repertoire. All the body strobing we had been seeing didn't make any sense if it was occurring in pitch darkness. However, since small photophores cover the Humboldt's body, including the mantle, head, tentacles, arms, and fins, both top and bottom, it seemed likely that the same body strobing could be replicated with bioluminescence but that we weren't seeing it because our red lights were too bright.

There *was* one dive where I *thought* I saw bioluminescence from the squid, but I wasn't sure. We had been sitting in the dark with all of Deep Rover's lights turned off, and I was looking down, watching a group of Humboldt squid far below me, just barely visible in the red lights of the Triton. These squid were strobing, but instead of the usual red-and-white whole-body color change, it looked like blue bioluminescence flashing on and off. It was very dim, and, knowing how obliging the human brain can be at filling in details based on what one *expects* to see, I wanted video confirmation. But Orla had been loath to turn out the lights when squid were present, for fear of missing a shot. If the squid were hanging around enough to be filmed under our red-light illuminators, that was her priority. Going

* Footnote censored due to profanity.

dark on the off chance that they might be emitting bioluminescence seemed too risky a bet.

But this time we were specifically going to try to film the bioluminescence of the squid, as well as of the bioluminescent plankton in the water, which we planned to stimulate with a SPLAT screen we had jury-rigged on the Triton. I was confident that the low-light BBC camera was going to provide spectacular imagery from the plankton— a sure thing. We also planned to spend some time on the bottom filming all those chicken fish, including their bioluminescence—another sure thing.

The plan was to launch at ten A.M. and descend straight to the bottom, stopping to film only if we encountered squid. On the seafloor, we would spend some time filming *Enypniastes*, then we would start up slowly, recording bioluminescent plankton at 50-meter (164-foot) depth intervals, all the time hoping to come across an aggregation of squid. We had eight hours.

Once again it was me and Toby in the Deep Rover, and Orla and Hugh in the Triton, with Alan Scott as their pilot. It had taken us a little over an hour to reach bottom, and although the goal was to keep the subs close to each other, we had lost sight of the Triton well before we settled down on the seafloor, at three thousand feet. We began to get concerned because Al was not responding to Toby's attempts to raise him on the through-water comms. Toby tried to pick the Triton up on our sonar display but had no luck. Next he called the surface to see if mission control could make contact. No joy there, either. The team on the ship was, however, able to get a position for the Triton from the USBL (ultrashort baseline) beacon. They could also see it on the echo sounder, showing it to be a few hundred feet off the bottom, which seemed strange.

At least now we had a direction in which to start looking. As we motored toward them, Toby kept trying the comms, and after many tries, we finally got a curt reply from Al saying that he was dealing with an intermittent failure of his computer control system. This was simultaneously a relief and a crushing disappointment—a relief because it was not life-threatening, since he could simply blow ballast and surface the sub without computer control, but disappointing because it meant we might have to abort the dive.

By the time we reached them, Al had managed to regain enough control that he had the sub sitting on the bottom, and as we drew closer I could see Orla in the starboard seat, her legs tucked under her, looking almost relaxed except for the sour expression on her face, in stark contrast to her usual cheery smile. While Al kept fighting with the computer, Toby and I reconnoitered the surrounding area. As we cruised over the bottom, I mused on the irrefutable nature of Murphy's law: Not only were we facing an equipment meltdown, but the ocean had thrown us another curve. There weren't any *Enypniastes* anywhere in sight! Instead of chicken fish, the seafloor was carpeted with large, bottom-dwelling shrimp, each resting in its own shallow indentation in the sediment, golden eyes staring up at us as we soared over them like an invading UFO.

It was apparent that our carefully laid filming plans were kaput. Al had manual override for some of the functions he needed, but trying to film the bioluminescence was proving to be an exercise in frustration. He had control of the thrusters, but he couldn't turn lights on and off as needed. Nevertheless, both subs remained cruising along the bottom looking for *Enypniastes* for nearly four hours and eventually found a few, but there was not the magnificent profusion we had seen on previous dives.

With only three hours of our eight-hour dive time left, the two submersibles began to move up through the water column looking for squid. Our expectations weren't high, because we had seen only *one* during our descent. But to keep the dive from being a total bust, whenever we saw anything slow-moving, like a jellyfish, that didn't require a lot of maneuvering by the Triton, we would stop and film it.

It was approaching six P.M. by the time we reached the oxygen minimum zone, and the batteries on both Triton and Deep Rover were nearly drained. We would have to surface soon. The Triton trailed below us, and while waiting for it to catch up, we were surprised by a Humboldt that shot up from below and attacked the underside of the e-jelly. I checked the depth—692 feet. We were smack-dab in the center of the oxygen minimum zone. It's known that Humboldt squid have a remarkable capacity for surviving very low oxygen by shutting down certain metabolic pathways, but it was assumed that they further reduced oxygen demand when in the OMZ by forgoing active hunting. Well . . . scratch that hypothesis. This was the very definition of active hunting.

That observation alone would have made the dive worthwhile for me, but the squid weren't done with their surprises. As we continued to move up, we came into a layer of lanternfish, still within the oxygen minimum zone, and, suddenly, more Humboldts than we had ever seen: hundreds of them, and they were actively hunting the fish. Although we had our white lights on, the squid seemed unperturbed, and, in fact, they were using our lights to see their prey. It was a frenzy of activity, and hard to take it all in. The Triton was below and off to our starboard side, filming the action, so all we had to do was sit still and illuminate the scene for them. And there certainly was no shortage of action. In any direction, they could point their cameras and see squid gliding through the water and hammering at prey.

The squid seemed to spend equal time swimming backward and forward. They would cruise backward, flapping their enormous fins, and then reverse on a dime when they spotted a target. I watched as one after another homed in on its prey, tentacles shooting out, sometimes adjusting their trajectory, even bending their tentacles at the last instant to intercept a fish as it zigged or zagged, sometimes missing, sometimes scoring a direct hit, pulling the fish back to where it disappeared inside a squid's splay of arms, and sometimes scoring a partial hit that resulted in a spray of shimmering scales.

They weren't infallible hunters, but they were persistent even as the fish were thinning out.

We had been there more than ten minutes, enough time to attract a large swarm of krill, drawn to our lights like moths to a flame. The krill were becoming so thick that they were making it hard to see the squid. Suddenly, a squid swimming directly toward us spread its arms and tentacles wide, forming a kind of basket, and then curled them back toward its mouth, scooping the krill in like popcorn. I had never seen squid feed like that. In fact, I didn't think anyone had. More squid started coming in then, basket-feeding on the krill.

I wanted to document this, but the Triton was too far away, filming the fishing action at the periphery of our light field. I had my Nikon, but at that precise moment it stopped shooting. I assumed the battery had died and changed it out, without taking my eyes off the show on display in front of the sub. More and more squid were swooping in, shoveling krill into their mouths like contestants in a shrimp-eating contest. When my camera still refused to shoot, I tore my eyes away long enough to realize that the problem wasn't the battery but a full memory card. As I dug around frantically for a spare, Toby pulled out his iPhone and managed to film thirty seconds, which included four basket-feeding attacks. Tens of thousands of dollars' worth of camera gear and what should have been the blue-chip natural history moment is shot on an iPhone! Fortunately, I found out later that the same feeding frenzy was happening around the Triton, so they were able to record some of that odd behavior as well.

We had been filming the squid for about fifteen minutes when suddenly they bolted, swimming explosively from left to right across our field of view. It was as if something spooked them—badly. Later, when we were back on the ship and talking to the crew, we learned that at about the time that the squid fled, a Chilean military helicopter had buzzed the ship, followed shortly afterwards by a high-speed military vessel that whizzed past at nearly twenty-four knots.* Presumably, the sound had alarmed the squid. Such an extreme flight response sure looked like predator avoidance behavior.

Squid don't have ears, but they do have statocysts, which allow them to detect low frequencies, below five hundred hertz. Toothed whales would be one sound-producing predator worth avoiding, since they can consume over two thousand pounds of squid each day. However, the only sounds they are known to generate are ultrasonic clicks that they use as a kind of biosonar to locate prey and to communicate. These clicks are composed of frequencies of seventeen thousand hertz, far above the range that squid can detect. In fact, it's recently been shown that squid are oblivious to and unharmed by ultrasound pulses broadcast at a decibel level that, if in our hearing range, would rupture human eardrums.

So if these squid are insensible to the sounds generated by such obvious predators as toothed whales, what could have elicited such a panic response? Well, another sound-producing predator worth avoiding is humans, since we are equally voracious consumers of squid. Have Humboldt squid learned or evolved to avoid engine noise? Obviously, these animals are highly adaptable. They seem to have a plethora of feeding strategies, shifting prey preference as the situation warrants—from fish to krill to each other. They can tolerate extremely low oxygen levels and, to some degree, may actually be beneficiaries of climate change. Their range in the eastern North Pacific Ocean has recently expanded, and they have invaded waters along the central California coast and been spotted as far north as the Gulf of Alaska. Their extreme adaptability positions them as potential survivors in a rapidly changing world, so it wouldn't surprise

---

*The Chilean navy was suspicious of our diving operation, despite the fact that we'd acquired all the necessary permits and had a Chilean scientist observer on board.

me at all to learn that they have developed a way to detect and avoid motorized fishing fleets.

Back on deck, I was practically levitating with excitement. *What fantastic, mysterious creatures these squid are!* We had been granted the oh-so-rare opportunity to observe them in their inner sanctum, and, as so often happens, the experience created more questions than it answered. I desperately wanted to know what had startled those squid! If it *was* sound, I wondered if noise produced by our support vessel had had any bearing on when we saw squid and when we didn't. Had I really seen them flashing bioluminescence that one time? What was all their body flashing meant to communicate? And how does the bioluminescence potential in the water impact their behavior? At that moment, I felt that I could happily spend the rest of my life studying just this one amazing patch of the sea.

GAINING A BETTER understanding of how this bit of ocean works relates directly to understanding how Spaceship Earth functions. Buckminster Fuller, a man of many titles, including inventor, architect, systems theorist, and futurist, popularized that phrase to emphasize what it means to live in a biological system with finite resources. If we damage our life-support machinery beyond repair, there is no possibility of a resupply ship showing up in the nick of time to save us. With that in mind, you might think that the importance of understanding how our world operates should be self-evident, but experience suggests otherwise. We humans have a really unfortunate history of not understanding the value of what we've got until it's gone. The collapse of fisheries around the world is just one of far too many examples.

One such fishery that I have direct experience with is Georges Bank, in the Gulf of Maine. Just seventy miles due east of Cape Cod, Georges Bank is an underwater plateau that is larger than the state of Massachusetts and was once a lush Garden of Eden, thanks to the confluence of two major ocean currents: the cold, nutrient-rich Labrador Current, sweeping down from the north, and the warmer Gulf Stream, coming up from the south. Where these ocean rivers meet, a profusion of plankton fed a rich ecosystem of marine life, including herring, cod, swordfish, haddock, yellowtail flounder, scallops, and lobsters, as well as more charismatic megafauna like dolphins, porpoises, turtles, whales, and seabirds.

North American Indians undoubtedly benefited from the ocean's bounty here, as did the seagoing Basques, from northern Spain, who claimed to have discovered these bountiful fishing grounds nearly half a century before Columbus supposedly discovered America. The history of this fishery follows those of fisheries around the world, going from waters so overflowing with life that early chroniclers described scooping the fish out of the water with baskets to overexploitation. As fishing stocks declined, modern fishing technology found ways to compensate, using aircraft and sonar to relentlessly track down schools of fish in the open ocean and going after bottom fish with massive trawling operations that decimated vital seafloor habitat. The use of enormous factory ships made it possible to haul in as much cod in a single hour as typical seventeenth-century vessels caught in a whole season (about one hundred tons). Although government agencies were advised that fishing stocks were being dangerously depleted, they nonetheless caved to short-term commercial fishing interests, until the fisheries on Georges Bank inevitably collapsed.

Most of us know the children's story of the goose that laid the golden egg: A peasant discovers a goose that lays one golden egg a day. Selling these eggs makes him rich, but the richer he grows, the greedier he becomes, until, in an attempt to extract all the gold at once, he slices the goose open—finding nothing while forever losing the source of his wealth. In the case of Georges Bank, the goose died in the early nineties, which led to a far-too-late fishing ban that was instituted at the end of 1994.

The presumption was that, given time, the fishery would recover.

But this presumption assumes that if you create a gaping hole in the web of life, it will restock with whatever was removed. Often, though, that niche is filled by something far less desirable. Ecosystems depend on feedback loops to maintain stability. If one or more of these feedback controls are radically altered, they become increasingly unstable, with the result that even small changes can produce very big effects. These are known as tipping points.

I was witness to what can happen when a tipping point tips, on my very first Johnson-Sea-Link dive, in 1989, which was in Wilkinson Basin, just north of Georges Bank. As soon as we entered the water, it was immediately obvious that this had become a jelly-dominated system. There were huge numbers of comb jellies (*Euplokamis* sp., *Pleurobrachia pileus*, *Bolinopsis infundibulum*) and siphonophores (*Nanomia cara*). This made for spectacular luminescence, but it also heralded a significant stumbling block to the recovery of the fishery, because not only do fish and jellyfish compete for plankton, but jellyfish consume fish eggs and fish larvae.

Multiple stressors contributed to the breakdown of the Georges Bank fishery. It wasn't just the removal of the fish. There was also the elimination of critical feedback loops like jellyfish predators, including leatherback turtles and swordfish. Nutrient runoff and sewage outflow from land produced regions of low oxygen that favored jellyfish over fish. Acidification, resulting from the ocean absorbing increasing amounts of carbon dioxide, drove down the pH, which was bad for fish but good for jellies. And changing temperatures and current patterns also favored jellyfish. The result was that the survival odds had so overwhelmingly tipped in favor of jellyfish over fish that reducing fishing pressure was no longer sufficient to return the system to balance. Had that complexity been understood earlier, it's possible that the fishing ban could have been implemented in time to save the goose.

This is why it's essential to invest time and money to explore and understand this patch of ocean off Peru and Chile, where Humboldt squid are the most heavily fished squid in the world. So far they seem to be holding their own. This is partly because it's a relatively new fishery. Squid used to be considered of low economic impor-tance* and were therefore left alone. It also benefits from being an artisanal fishery where the primary method of harvesting the squid is jigging—catching squid on hooks—which helps limit not only the catch but also the bycatch, all the recklessly wasted marine life that is caught with less selective fishing gear (like nets) and then simply discarded back into the ocean.

Historically, governments have allocated serious money toward studying ecosystems only *after* they collapse (and often not even then). This occurs in response to the populace screaming, "Fix it! Put it back the way it was!" But how can we possibly do that if we didn't study it while it was working? We haven't even adequately *ex-plored* the ocean, let alone carried out the kind of long-term studies and observations needed to develop a truly useful operating manual, never mind a *repair* manual.

In the meantime, humans aren't just pressing gently on the com-plex levers, gears, and switches that control the planet's life-support machinery—we are *jumping up and down on them,* with about as much forethought as children bouncing on a waterbed. It's been fun while it lasted, but things are beginning to go catastrophically wrong.

In response, some people have adjusted their perspective away from what seem to be insurmountable problems here on Earth, or, at the very least, problems that are out of their control, to focus on space exploration. Our need to explore is so inherent that we em-brace space exploration on the very slimmest of rationales. The music soars, the rockets thunder, and the narrator opines, *It is our destiny to explore the cosmos . . . We need space exploration in order to*

---

*This is what's known as fishing down the food web, where, once the most desir-able fish are eliminated, increasingly less desirable species are fished out until we're all eating jellyfish stew.

inspire the next generation of explorers . . . It serves to stimulate the public's imagination . . . We need to study other planets to better understand Earth . . . Explorers are us. These points are all true, but are they rational, given the increasingly hot water surrounding us?

Our focus needs to be on exploring our own planet before it's too late. We know that our oceans are what make our planet livable, and yet they remain mostly unknown. We need to launch a new age of exploration, one that is focused on our greatest treasure, *life.*

As far as we know right now, Earth is unique in its ability to sustain life. How it manages this miracle is still largely a mystery—one that seems worthy of further examination if we hope to go on enjoying its benefits. I embrace exploration—all exploration—because there is always new knowledge to be gained. But in the face of limited budgets and hard choices, I turn my head away from the stars to look instead at our oceans. I choose life and our own existence, as well as that of swaying kelp forests with playful sea otters and neon-orange garibaldi fish; seagrass meadows with laconic grazing manatees, fantastic leafy sea dragons, and columns of spiny lobsters on parade; coral reefs composed of elaborate, interwoven poster-colored living architectures surrounded by a swirling kaleidoscope of glittering gems like electric-blue damselfish, lemon-yellow tangs, and opal-hued parrotfish; translucent blue waters swarming with planktonic life and harboring vast shoals of silvery fish wheeling and plunging alongside titanic blue whales breeching and dolphins leaping and twirling, whistling and clicking; icebergs drifting in the sunlight, providing respite for polar bears, walruses, and puffins; and, of course, deep-sea coral gardens awash in twinkling bioluminescent splendor and patrolled by giant sixgill sharks and giant squid drifting at the edge of darkness. I know I'm biased, but seriously, how can the barren surface of Mars possibly compare?

Many of these amazing natural wonders are disappearing on our watch, long before we've been able to suss out the intricacies of their existence in the universe. As we strip the natural world of all its bounty, slaughtering one golden goose after another in the name of get-rich-quick schemes, we are complicit in the ultimate Ponzi scheme—kicking the can down the road for our progeny to deal with. In the face of this decimation, the notion that we should be focusing our time and resources on visiting lifeless rocks is so patently absurd that future generations struggling to survive among collapsing ecosystems and an unstable climate may reasonably ask, "What the hell were you thinking?"

Most people recognize that the drive for exploration is in our DNA, but it behooves us to examine what purpose that drive serves. For primitive humans, gaining knowledge about something, like which mushrooms are poisonous, had enormous value. Knowledge was and is our most valuable resource. It is the thing, more than any other, that sets us apart as a species. We assimilate knowledge both collectively and cumulatively.

Collective knowledge means that not everyone in a community needs to be able to recognize which mushrooms are poisonous, as long as at least one trusted member of the tribe has that knowledge and is willing to share it. And cumulative knowledge means that lessons about which mushrooms are poisonous don't need to be relearned the hard way by each succeeding generation, because we have found ways to accumulate such data and pass it on, first through an oral tradition and later through the written word and now the World Wide Web.

For a long time, the fact that bad information—like *Earthquakes are caused by sinners*—could be passed on as easily as good information—like *Don't eat the death caps!*—slowed progress. But once the scientific method was developed and provided a means to test for truth, knowledge accumulated exponentially, and civilization flourished.

Our ever-expanding ability to share information sets humans apart from any other species on our planet. This ability has fostered a different, faster kind of evolution than that initially produced by

Darwinian natural selection. Known as human cultural evolution, it is a process that is not constrained by the random genetic variations that selection acts upon. The staggering power of this new kind of evolution stems from the fact that all that is learned in one generation can be passed directly on to the next. That doesn't happen in any other species.

What is astonishing about this explosion of human communication, which far and away exceeds any other animal signaling system on the planet, is that we can't manage to communicate the most important information of our time: that we are decimating the natural world and, in so doing, imperiling our own existence.

Our survival instincts are failing us because the threats we are facing are not the kind we evolved to perceive and respond to. If we don't *see* the danger, we often have a hard time believing it's real—something made painfully clear by the range of maladaptive responses exhibited to the danger posed by Covid-19. The conclusion of many of those studying the response to the pandemic, as well as to climate change, is that knowledge without feeling is an inadequate motivator.

For some people, using fear as a reason to act works, but for many others the danger feels too remote. Our problem is how to reconcile the danger presented by the impending, but not quite immediate, doom of climate change with the proximate problem of missing the mortgage payment or paying the cable bill. In the face of such different demands, the one that wins out is the one that feels pressing but is ultimately far less important. To turn this around, I think we need to adjust our viewpoint by focusing on the positive rather than the negative. This is because, as any parenting guide will tell you, nagging and cajoling your offspring into doing something they don't want to do is usually counterproductive. Positive motivators work better.

According to child developmental psychologist Alison Gopnik, "making it fun" is also how nature gets us to do things that are good for us. In fact, she makes the provocative claim that, just as orgasm provides positive reinforcement for making babies, the feelings of surprise and joy that are associated with figuring out how the world works are just as powerful. Now there's an idea worth spreading: *Exploring is better than sex: an equivalent high without the baggage!*

If you've ever tried to "childproof" a house against the exploratory instincts of a two-year-old, you know what a powerful drive she's talking about. According to Gopnik, feelings of intense curiosity and surprise are so overwhelming in young children that "they put themselves at mortal peril for the sake of figuring out the causal structure of the world."* From my own personal experience and those of many of my colleagues, I can attest that the intensity of those emotions do not diminish with age. They are there for everyone to tap into, and right now we need them more than ever.

We *will* venture into the deep ocean, because that's what humans do: We explore. Even as government funding for ocean exploration has continued to shrink, private funding is on the rise, and new deep-diving technologies are being developed. In 2012, film director and deep-sea explorer James Cameron privately funded the development of the single-person submersible Deepsea Challenger, which he dove to the deepest point in the ocean, in the Mariana Trench off Guam (35,756 feet). In 2018, wealthy American businessman Victor Vescovo launched the Five Deeps Expedition, diving to the deepest points of the five world oceans in a submersible called the DSV *Limiting Factor,* which he commissioned to be built by Triton Submarines. And in 2020, billionaire philanthropist Ray Dalio launched the 286-foot research vessel *OceanXplorer,* with three submarines, two undersea robots, a helicopter, onboard wet and dry labs, livestreaming and video production capabilities, and a goal of con-

---

*Courtney Stephens and Alison Gopnik, "Why Ask Why? An Interview with Alison Gopnik," *Cabinet Magazine,* Fall 2005, cabinetmagazine.org/issues/19/stephens.php.

necting people with what he calls "our world's greatest asset." I love the fact that in describing the wonders to be revealed, Dalio related an experience diving in the deep Pacific in 2013: He was in pitch darkness when a camera flashed and stimulated waves of bioluminescence from the surrounding creatures. "It was like a fireworks display," he said. "Everything was responding. It was unbelievable."*

The question is: Have we learned enough from history about the shortsightedness of exploration simply for exploitation? We need to be smarter about how we approach our planet's last frontier and our own future, which means fully embracing the certainty that our most precious resource is not oil or metals—it's life.

Humans have an amazing capacity for changing our perspective. We can zero in on the inner workings of a cell or the mechanics of subatomic particles, or we can expand our field of view to imagine an infinite cosmos. It is our capacity to adjust focus as needed that is our superpower. Right now, our future on the planet depends on concentrating on what makes life possible, which means we need to be able to see life with new eyes.

Bioluminescence makes life that was once invisible and obscure brilliantly observable. In the deep and vast darkness of the largest living space on the planet, a single, tiny flash announces the extraordinary experiment that is life. It is utterly remarkable, and too little appreciated, that a creature as small as a dinoflagellate—less than forty microns in diameter and invisible without a microscope—can be seen by its flash several feet away! In fact, most of the animals in the ocean are likewise self-tagged with very distinctive forms of bioluminescence, providing us with a means to view life as never before. If the purpose of life is to understand itself, then perhaps living light can help illuminate the path to that destination.

---

*William J. Broad, "A New Ship's Mission: Let the Deep Sea Be Seen," *New York Times,* September 17, 2020.

A CASE FOR OPTIMISM

These days, you need to be an optimist if you are going to be an environmentalist. You *must* believe you can make a difference. Otherwise, why bother?

In 2005, I helped found an environmental nonprofit, the Ocean Research & Conservation Association (ORCA). As the submersible program at Harbor Branch began winding down, I realized I needed to focus on some increasingly pressing concerns related to the health of the ocean. Reports from two monumental studies* had recently grabbed my attention. The consensus was that the ocean was in crisis. To address the danger, both reports emphasized a great need for more advanced monitoring technologies.

In a hospital, everyone appreciates the importance of monitoring. If you go to the emergency room with an undiagnosed illness, the first thing the doctors do is start monitoring your various life-support systems—your heart, your blood, and your lungs. They need to do this to figure out what's wrong and then determine whether their treatment of choice is making you better and not worse. We need to do the same for our planetary life-support systems.

---

*The Pew Oceans Commission, in 2003, and the U.S. Commission on Ocean Policy, in 2004.

We also need more monitoring to provide improved forecasting. As our climate becomes increasingly unpredictable, better forecasting of sea level rise, storm surge, floods, tsunamis, and hurricanes will save untold lives, and the ability to better pinpoint whom to evacuate and where to deploy sandbags will save beaucoup dollars. This is what's known as "environmental intelligence," and in a rapidly changing world, it is one of the most cost-effective investments that any nation can make in its future economy and security.

In starting ORCA, my thinking was to focus on developing technological solutions to collect that environmental intelligence, with improved cost-effective monitoring being the primary focus. The Moored Eye-in-the-Sea fit neatly under that umbrella, as did the high-tech coastal water-quality monitors that we were developing, called Kilroys (so named because we hope they will eventually be everywhere, like that ubiquitous GI of World War II fame). As conceived, the Kilroys were small, solar-powered systems that were attached to moorings and used cellphone technology to communicate a wide range of water and pollution parameters to the internet. Normally if you build something half the size and at two-thirds the cost of comparable systems, you should be golden, but our momentum was slowed by the economic downturn in 2008. With state and federal monitoring programs being slashed to the bone, my business plan, which depended on selling Kilroys to finance conservation science, was made null and void. I needed a plan B.

My goal with the Kilroys was to track pollution to its source and find ways to stop it, so I started casting about for even less expensive ways to accomplish the same thing. It set me to thinking about how living organisms were being impacted by pollutants entering our local waters.

I have the great good fortune to live on the shores of the Indian River Lagoon, a 156-mile-long estuary along the eastern seaboard of Florida. Once called the most biologically diverse estuary in the United States, it is a shallow-water lagoon bordered by lush mangrove forests and inhabited by amazing wildlife. When David and I first moved here, in 1989, we would routinely see roseate spoonbills flying over our house every morning, manatees would come up to our dock and drink from our hose, otters would climb onto our dock and use our dock pilings to scratch their backs, and, at sunset, the sound of mullet jumping and slapping the water would fill the air.

Estuaries are the nurseries of the ocean, where many open-ocean inhabitants come to spawn because of abundant food and hiding places among habitats like mangrove roots and seagrass meadows. They also provide sustenance for terrestrial wildlife, including many year-round and migrating birds. As a result, estuaries are among the most biodiverse habitats on the planet—right up there with coral reefs and rainforests. If you are going to pick a key ocean ecosystem to focus on protecting, this is it.

The wildlife sightings that we reveled in when we first moved here are now vanishingly rare. Water quality has declined, much of the seagrass has disappeared, and we see dolphins with cauliflower-like fungal lesions called lobomycosis and sea turtles with huge, debilitating fibropapilloma tumors.

The estuary is being poisoned by pollutants from agriculture and lawns running off the land as well as leaky sewage and septic systems. I wanted to pinpoint where the toxicants were coming from. Testing for specific chemicals can be very expensive, especially if you have no clue what to test for, so I started looking for a living system that could be used as an indicator, a canary in the coal mine, if you will. This is what's known as a bioassay and, not too surprisingly, the one I settled on was bioluminescent bacteria. I had begun to investigate a bioassay already on the market, called Microtox. It was based on harmless bioluminescent bacteria. Because their light output is linked to their respiratory chain, any toxicant that poisons respiration, as most do, interferes with light output. I wanted to use Microtox to test for toxicity in sediment samples taken from the bottom of the estuary. Several investigators had tried to do this already, but it had proven unreliable and was largely abandoned. To meet the challenge, I hired a scientist, Beth Falls, who has a true explorer's can-do approach, and she found a way to make it work.

Because pollutants often persist in sediment much longer than in water, this bioluminescence assay allowed us to locate pollution sinks—the places where pollution was most concentrated in the estuary. In order to share our findings with the public and with policy makers, ORCA began producing pollution maps. Our maps look just like weather maps, on which red is hot and blue is cold, but on ours, red is toxic and blue is nontoxic. These maps don't tell us what the toxicants are, but they do tell us where to focus our sampling and mitigation efforts. This approach saves huge amounts of time and money. So much so that we have now greatly expanded the methodology, sampling sediments for a whole range of pollutants, including nutrients, and using maps to measure the impact of mitigation projects. Our tagline for ORCA, "Mapping Pollution, Finding Solutions," summarizes that approach.

We have been fortunate enough to get a lot of community support for these efforts, initially working with local high school students to collect and analyze the data and, more recently, with ORCA-trained citizen scientists. Getting local citizens involved is a force multiplier; it greatly expands the amount of data we can collect, and it helps create powerful, knowledgeable advocates for our precious estuary. Just as a well-informed electorate is a prerequisite for democracy, a science-literate citizenry is essential for maintaining the health of the planet.

But in talking with that citizenry, as well as with my own incredibly hardworking team at ORCA, I am aware of how much eco-anxiety we are up against. This is a widespread but counterproductive suite of emotions that can cause people to tune out and give up. As a result, over the years I have developed a talk on optimism, in which I start out by joking that I am not an optimist by birth, but by marriage. I describe how my husband, David, is one of the most optimistic people I've ever met. In fact, when we were first married, I thought he didn't have a very firm grip on reality, because sometimes his optimism seemed to defy logic, but he is not a Pollyanna. He believes in preparing for the worst but expecting the best.

What turned me into a convert was the number of times over the years that I have seen him locate a pony in a heaping pile of manure just because he was absolutely convinced it had to be in there. It is only the optimists who will find the solutions, so I have been saying for a while now that we need to stop preaching hopelessness and instead focus on empowering the next generation of explorers with the tools they will need to find those ponies.

To clarify what kind of optimism I mean, I used to talk about what's called the Stockdale Paradox. This was a story that came out of the business book *Good to Great*, by James Collins, in which he shared an interview he did with Admiral James Stockdale, who had been a prisoner at the infamous "Hanoi Hilton" POW camp during the Vietnam War. Seven and a half years of abject misery and unspeakable torture did not break this man, who managed to maintain not only his own morale but that of other prisoners in the camp. When asked about his coping strategy, Stockdale said, "I never lost faith in the end of the story, I never doubted not only that I would get out, but also that I would prevail in the end and turn the experience into the defining event of my life, which, in retrospect, I would not trade."

Asked about the prisoners who were broken by the experience and didn't make it, he said, "Oh, that's easy, the optimists. Oh, they were the ones who said, 'We're going to be out by Christmas.' And Christmas would come, and Christmas would go. Then they'd say, 'We're going to be out by Easter.' And Easter would come, and Easter would go. And then Thanksgiving, and then it would be Christmas again. And they died of a broken heart. This is a very important lesson. You must never confuse faith that you will prevail in the end—which you can never afford to lose—with the discipline to confront the most brutal facts of your current reality, whatever they might be." The duality of Stockdale's capacity to balance the merciless reality of his seemingly hopeless situation with the unwavering belief that he would eventually triumph is why this is called the Stockdale Paradox. It is a form of optimism, but one that many people find confounding.

In recent years, I have found it easier to convey the kind of optimism I'm talking about by referring my audience to Matt Damon's character Mark Watney in *The Martian*. Yes, I totally get the irony of that, but it's a fantastic book and a brilliant movie that completely captures the ethos of what it means to be an explorer. Although Watney's situation—presumed dead and abandoned on Mars—seems beyond hopeless, he doesn't delude himself about his circumstances. He confronts his reality and just keeps working the problem, taking on challenges in the order of their priority.

And so, to close, let me leave you with these two thoughts:

Optimism is a thing worth fighting for. We have to keep trying and never give up the faith that we will prevail.

And, to paraphrase Mark Watney, in the face of overwhelming odds, we're left with only one option: We're going to have to "science the shit out of this."

ACKNOWLEDGMENTS

A LONG, LONG time ago, in 1997, I wrote a book called *Light Soup,* which I intended as an introduction to bioluminescence. I found an agent, who shopped it around to various publishers, who all said the same thing: "You need to make it more personal." As a scientist, trained to never write in the first person, I had no idea how to do that, so I put it on a shelf.

It was December 2011 when literary agent Farley Chase, having seen an article about my research in *The New York Times,* tracked me down to ask if I had ever considered writing a memoir. My answer, at the time, was a firm no.

A year and a half later, after the giant squid documentary aired, he contacted me again. He made a compelling case for the things I had seen and the stories I could tell. This time I sent him my *Light Soup* manuscript. He managed to be laudatory while at the same time saying what I had already heard: "You need to make it more personal." My answer was still "No. I don't know how to do that."

His gentle persistence is evident in the more than forty emails that followed, in which he provided encouragement and reading recommendations until mid-2015, when he finally convinced me to give it a try. "Trying" involved two years of sending him outlines and sample chapters, until early 2017, when I finally had something that he thought was good enough to put into a book proposal.

In other words, there is no way this book would have happened without Farley Chase, so he gets top billing.

It was Farley and especially my phenomenal editor at Random

House, Annie Chagnot, who patiently guided me through the writing process with lots and lots of admonitions to "show, don't tell!" She managed to provide just the right mix of critique and encouragement. Annie deserves my heartfelt gratitude and admiration for helping me to rethink the first draft and for working so hard on knocking off the revised second draft chapters in heroic fashion during the pandemic and her pregnancy, practically editing on her way to the delivery room—where she delivered a healthy baby girl.

My wonderful husband, David, read every draft and provided lots of helpful feedback, bestowed plenty of encouragement when required, and helped me carve out the extra time I needed for writing by taking over my household chores, including all the cooking. That was his idea, not mine, and although I was initially hesitant, because his cooking had previously been pretty hit-or-miss (one might say overly optimistic), he has evolved into an excellent chef. When I was asked in an interview recently what my smartest career decision was, I said without hesitation, "Marrying my husband." He is the wind beneath my wings and the water beneath my fins.

Other book-related help along the way came from Julie Grau, who imparted valuable advice on the early chapters; associate editor Rose Fox, who ably filled in while Annie Chagnot was on maternity leave; copy editor Will Palmer, who clearly put a lot of thought into getting the wording just right; Tammy Frank, who provided valuable science feedback; and Richard Dawkins, who contributed insightful comments on the first four chapters.

For helping to hone my literary chops as well as providing encouragement related to an early draft, when I'm not at all sure it was warranted, I thank the wonderful gals of my book club, Robin Dannahower, PJ Dempsey, Jan Fehrman (who is deeply missed), Michelle Lineal, Leigh Hoppe, Sue Van Dyke, and Wendy Williams.

Special thanks to U.S. Navy Rear Admiral Thomas Q. Donaldson V, who provided the quote from German U-boat commander Captain Reinhard Hardegen.

Most of the stories told here were about large collaborative ventures, which means many, many names were left out. To all those collaborators, sub crew, ship's crew, colleagues, and friends who played a vital role, thank you. There are too many names to list here, but please allow me to share just a few that got left out of these pages but made a critical difference at key times. If I haven't said thank you before, I'm saying it now: Mel Briscoe, Mary Chapman, Tony Cimaglia, Andrew Clark, Larry Clark, Dave Cook, Jerry Corsaut, Jim Eckman, Warren Falls, Marjorie Findlay, Herb Fitz Gibbon, Geoffrey Freeman, Steve Haddock, John Hanke, Peter Herring, Page Hiller-Adams, George Jones, Patrick Lahey, Janeen Mason, Edwin Massey, Gene Massion, Harry Meserve, Milbry Polk, Eric Reese, Vin Ryan, Mark Schrope, Chris Tietze, D. R. Widder, and Charlie Yentsch.

FURTHER READING

Introduction: A Different Light

Barbier, Edward, et al. "Ecology: Protect the Deep Sea." *Nature* 505, no. 7484 (2014): 475–77.

Danovaro, Roberto, Cinzia Corinaldesi, Antonio Dell'Anno, and Paul V. R. Snelgrove. "The Deep-Sea Under Global Change." *Current Biology* 27, no. 11 (June 2017): R461–65. www.sciencedirect.com/science/article/pii/S0960982217302178.

Drazen, Jeffrey C., et al. "Opinion: Midwater Ecosystems Must Be Considered When Evaluating Environmental Risks of Deep-Sea Mining." *Proceedings of the National Academy of Sciences* 117, no. 30 (July 2020): 17455–60.

Nouvian, Claire. *The Deep: The Extraordinary Creatures of the Abyss.* Chicago: University of Chicago Press, 2007.

Ramirez-Llodra, Eva, et al. "Deep, Diverse and Definitely Different: Unique Attributes of the World's Largest Ecosystem." *Biogeosciences* 7 (2010): 2851–99.

RealClimate: Climate Science from Climate Scientists. www.realclimate.org.

Chapter 1: Seeing

Cronin, Thomas W., Sönke Johnsen, N. Justin Marshall, and Eric J. Warrant. *Visual Ecology.* Princeton, N.J.: Princeton University Press, 2014.

Johnsen, Sönke. *The Optics of Life: A Biologist's Guide to Light in Nature.* Princeton, N.J.: Princeton University Press, 2012.

Purves, Dale, R. Beau Lotto, and Surajit Nundy. "Why We See What We Do." *American Scientist* 90, no. 3 (January 2002): 236–43. www.americanscientist.org/sites/americanscientist.org/files/20051220143043_306.pdf.

Chapter 2: Fiat Lux

Fleiss, A., and K. S. Sarkisyan. "A Brief Review of Bioluminescent Systems." *Current Genetics* 65, no. 4 (August 2019): 877–82.

Haddock, S. H. D., M. A. Moline, and J. F. Case. "Bioluminescence in the Sea." *Annual Review of Marine Science* 2 (January 2010): 443–93.

Shimomura, O., and I. Yampolsky. *Bioluminescence: Chemical Principles and Methods.* 3rd ed. Hackensack, N.J.: World Scientific, 2019.

Wilson, Therese, and J. Woodward Hastings. *Bioluminescence: Living Lights, Lights for Living.* Cambridge, Mass.: Harvard University Press, 2013.

Chapter 3: First Flash

Cusick, K. D., and E. A. Widder. "Bioluminescence and Toxicity as Driving Factors in Harmful Algal Blooms: Ecological Functions and Genetic Variability." *Harmful Algae* 98 (September 2020): 101850. www.sciencedirect.com/science/article/abs/pii/S1568988320301293.

———. "Intensity Differences in Bioluminescent Dinoflagellates Impact Foraging Efficiency in a Nocturnal Predator." *Bulletin of Marine Science* 90, no. 3 (July 2014): 797–811.

Eckert, Roger, and Takao Sibaoka. "The Flash-Triggering Action Potential of the Luminescent Dinoflagellate *Noctiluca.*" *Journal of General Physiology* 52, no. 2 (1968): 258–82.

Hanley, K. A., and E. A. Widder. "Evidence That the Adaptive Value of Bioluminescence in Dinoflagellates Is Concentration Dependent." *Photochemistry and Photobiology* 93, no. 2 (March/April 2017): 519–30.

Harvey, E. Newton. *A History of Luminescence: From the Earliest Times Until 1900.* Mineola, N.Y.: Dover, 2005.

Hoppenrath, Mona, and Juan F. Saldarriaga. "Dinoflagellates." Tree of Life Web Project. Version 15, December 2012. tolweb.org/Dinoflagellates/2445/2012 .12.15.

Widder, E. A., and J. F. Case. "Bioluminescence Excitation in a Dinoflagellate." In *Bioluminescence: Current Perspectives,* edited by K. H. Nealson, 125–32. Minneapolis: Burgess, 1980.

———. "Distribution of Subcellular Bioluminescent Sources in a Dinoflagellate, *Pyrocystis fusiformis.*" *Biological Bulletin* 162, no. 3 (June 1982): 423–48.

———. "Luminescent Microsource Activity in Bioluminescence of the Dinoflagellate, *Pyrocystis fusiformis.*" *Journal of Comparative Physiology* 145, no. 4 (1982): 517–27.

———. "Two Flash Forms in the Bioluminescent Dinoflagellate, *Pyrocystis fusiformis.*" *Journal of Comparative Physiology* 143, no. 1 (1981): 43–52.

Chapter 4: The Stars Below

Beebe, William. *Half Mile Down.* New York: Harcourt, Brace, 1934.

Herring, Peter. *The Biology of the Deep Ocean.* New York: Oxford University Press, 2002.

Rossotti, Hazel. *Colour: Why the World Isn't Grey.* Princeton, N.J.: Princeton University Press, 1985.

Widder, E. A. "Bioluminescence." In *Adaptive Mechanisms in the Ecology of Vision,* edited by S. N. Archer, M. B. A. Djamgoz, E. Loew, J. C. Partridge, and S. Vallerga, 555–81. Dordrecht, Netherlands: Kluwer Academic, 1999.

————. "Bioluminescence in the Ocean: Origins of Biological, Chemical, and Ecological Diversity." *Science* 328, no. 5979 (2010): 704–8.

————. "Marine Bioluminescence." *Bioscience Explained* 1, no. 1 (2001). www .medarbetarportalen.gu.se/digitalAssets/1566/1566428_biolumen.pdf.

Widder, E. A., M. I. Latz, and J. F. Case. "Marine Bioluminescence Spectra Measured with an Optical Multichannel Detection System." *Biological Bulletin* 165, no. 3 (December 1983): 791–810.

Chapter 5: Strange Illumination

Frank, T. M., and E. A. Widder. "The Correlation of Downwelling Irradiance and Staggered Vertical Migration Patterns of Zooplankton in Wilkinson Basin, Gulf of Maine." *Journal of Plankton Research* 19, no. 12 (December 1997): 1975–91.

Johnsen, S., E. A. Widder, and C. D. Mobley. "Propagation and Perception of Bioluminescence: Factors Affecting Counterillumination as a Cryptic Strategy." *Biological Bulletin* 207, no. 1 (August 2004): 1–16.

Marshall, N. B. *Developments in Deep-Sea Biology*. Poole, Dorset: Blandford, 1979.

Robison, B. H. "Midwater Biological Research with the WASP ADS." *Marine Technology Society Journal* 17, no. 3 (Fall 1983): 21–27.

————. "Running the Gauntlet: Assessing the Threats of Vertical Migrators." *Frontiers in Marine Science* 7 (February 2020): 1–10. www.frontiersin.org/articles/10.3389/fmars.2020.00064/full.

Tinbergen, Niko. *The Animal in Its World*. Cambridge, Mass.: Harvard University Press, 1972.

Widder, E. A. "Bioluminescence and the Pelagic Visual Environment." *Marine and Freshwater Behaviour and Physiology* 35, no. 1–2 (2002): 1–26.

Widder, E. A., and T. M. Frank. "The Speed of an Isolume: A Shrimp's Eye View." *Marine Biology* 138, no. 4 (April 2001): 669–77.

Chapter 6: A Bioluminescent Minefield

Widder, E. A. "SPLAT Cam: Mapping Plankton Distributions with Bioluminescent Road-Kill." In vol. 3 of *Oceans 2002 MTS/IEEE*, 1711–15. Piscataway, N.J.: Institute of Electrical and Electronics Engineers, 2002.

Widder, E. A., S. A. Bernstein, D. F. Bracher, J. F. Case, K. R. Reisenbichler, J. J. Torres, and B. H. Robison. "Bioluminescence in Monterey Submarine Canyon: Image Analysis of Video Recordings from a Midwater Submersible." *Marine Biology* 100, no. 4 (1989): 541–51.

Widder, E. A., and S. Johnsen. "3D Spatial Point Patterns of Bioluminescent Plankton: A Map of the 'Minefield.'" *Journal of Plankton Research* 22, no. 3 (2000): 409–20.

Chapter 7: Seas Sowed with Fire

Case, J. F., E. A. Widder, S. A. Bernstein, K. Ferer, D. Young, M. Latz, M. Geiger, and D. Lapota. "Assessment of Marine Bioluminescence." _Naval Research Reviews_ 45, no. 2 (1993): 31–41.

Lewis, David. _We, the Navigators: The Ancient Art of Landfinding in the Pacific._ Honolulu: University of Hawaii Press, 1972.

Rohr, J., M. I. Latz, S. Fallon, J. C. Nauen, and E. Hendricks. "Experimental Approaches Towards Interpreting Dolphin-Stimulated Bioluminescence." _Journal of Experimental Biology_ 201, no. 9 (1998): 1447–60.

Sontag, Sherry, Christopher Drew, and Annette Lawrence Drew. _Blind Man's Bluff: The Untold Story of American Submarine Espionage._ New York: PublicAffairs, 1998.

Vacquié-Garcia, J., F. Royer, A. C. Dragon, M. Viviant, F. Bailleul, and C. Guinet. "Foraging in the Darkness of the Southern Ocean: Influence of Bioluminescence on a Deep Diving Predator." _PLOS One_ 7 no. 8 (2012): e43565. doi.org/10.1371/journal.pone.0043565.

Widder, E. A. "Bioluminescence—Shedding Some Light on Plankton Distribution Patterns." _Sea Technology,_ March 1997: 33–39.

———. "A Look Back at Quantifying Oceanic Bioluminescence: Seeing the Light, Flashes of Insight and Other Bad Puns." _Marine Technology Society Journal_ 40, no. 2 (2006): 136–37.

Widder, E. A., J. F. Case, S. A. Bernstein, S. MacIntyre, M. R. Lowenstine, M. R. Bowlby, and D. P. Cook. "A New Large Volume Bioluminescence Bathyphotometer with Defined Turbulence Excitation." _Deep-Sea Research_ 40, no. 3 (1993): 607–27.

Widder, E. A., C. L. Frey, and J. R. Bowers. "Improved Bioluminescence Measurement Instrument." _Sea Technology,_ February 2005, 10–16.

Widder, E. A., S. Johnsen, S. A. Bernstein, J. F. Case, and D. J. Neilson. "Thin Layers of Bioluminescent Copepods Found at Density Discontinuities in the Water Column." _Marine Biology_ 134, no. 3 (August 1999): 429–37.

Wikipedia. S.v. "_Dolphins_ (M. C. Escher)." Last modified June 19, 2020. en.wikipedia.org/w/index.php?title=Dolphins_(M._C._Escher).

Chapter 8: Glorious Puzzles

Claes, J. M., D.-E. Nilsson, J. Mallefet, and N. Straube. "The Presence of Lateral Photophores Correlates with Increased Speciation in Deep-Sea Bioluminescent Sharks." _Royal Society Open Science_ 2, no. 7 (2015): 150219.

Davis, M. P., N. I. Holcroft, E. O. Wiley, et al. "Species-Specific Bioluminescence Facilitates Speciation in the Deep Sea." _Marine Biology_ 161 (2014): 1139–48.

Ellis, E. A., and T. H. Oakley. "High Rates of Species Accumulation in Animals with Bioluminescent Courtship Displays." _Current Biology_ 26, no. 14 (2016): 1916–21.

Haddock, S. H. D., M. A. Moline, and J. F. Case. "Bioluminescence in the Sea." _Annual Review of Marine Science_ 2, no. 1 (2010): 443–93.

Herring, Peter J., ed. *Bioluminescence in Action*. London: Academic Press, 1978.

Herring, P. J. "Sex with the Lights On? A Review of Bioluminescent Sexual Dimorphism in the Sea." *Journal of the Marine Biological Association of the United Kingdom* 87, no. 4 (2007): 829–42.

Johnsen, S., E. J. Balser, and E. A. Widder. "Modified Suckers as Light Organs in a Deep-Sea Octopod." *Nature* 398, no. 6723 (March 1999): 113–14.

Widder, E. A. "A Predatory Use of Counterillumination by the Squaloid Shark, *Isistius brasiliensis*." *Environmental Biology of Fishes* 53, no. 3 (1998): 267–73.

Wikipedia. S.v. "Johnson Sea Link." Last modified April 2, 2020. en.wikipedia.org/wiki/Johnson_Sea_Link.

Chapter 9: Stories in the Dark

Link, Marion Clayton. *Windows in the Sea*. Washington, D.C.: Smithsonian Institution Press, 1973.

Olson, Randy. *Don't Be Such a Scientist: Talking Substance in an Age of Style*. Washington, D.C.: Island Press, 2010.

Thaler, Andrew David. "The Politics of Fake Documentaries." *Slate*, August 31, 2016. slate.com/technology/2016/08/the-lasting-damage-of-fake-documentaries-like-mermaids-the-body-found.html.

Chapter 10: Plan B

Kaharl, Victoria A. *Water Baby: The Story of* Alvin. New York: Oxford University Press, 1990.

Kunzig, Robert. *Mapping the Deep: The Extraordinary Story of Ocean Science*. London: W. W. Norton, 2000.

Monterey Bay Aquarium Research Institute. "History." Accessed October 28, 2020. www.mbari.org/about/history/.

Raymond, E. H., and E. A. Widder. "Behavioral Responses of Two Deep-Sea Fishes to Red, Far-Red and White Light." *Marine Ecology Progress Series* 350 (2007): 291–98.

Widder, E. A., B. H. Robison, K. R. Reisenbichler, and S. H. D. Haddock. "Using Red Light for In Situ Observations of Deep-Sea Fishes." *Deep-Sea Research* 52, no. 11 (November 2005): 2077–85.

Wikipedia. S.v. "DSV *Alvin*." Last modified August 9, 2020. en.wikipedia.org/w/index.php?title=DSV_Alvin&oldid=971984201.

Chapter 11: The Language of Light

Herring, P. J., and E. A. Widder. "Bioluminescence of Coronate Medusae." *Marine Biology* 146, no. 1 (December 2004): 39–51.

"A More Candid Underwater Camera." *Popular Mechanics*, October 1, 2009. www.popularmechanics.com/science/environment/a3336/2628296/.

NOAA Ocean Explorer. "Operation Deep Scope 2004." Accessed October 28, 2020. oceanexplorer.noaa.gov/explorations/04deepscope/background/plan/plan .html.

———. "Operation Deep Scope 2005." Accessed October 28, 2020. oceanexplorer .noaa.gov/explorations/05deepscope/background/eyeinsea/eyeinsea.html.

Widder, Edith. "Glowing Life in an Underwater World." Filmed April 2010 on Mission Blue Voyage, Galápagos Islands, Ecuador. TED video, 17:05. www.ted.com/ talks/edith_widder_glowing_life_in_an_underwater_world?language=en#t -695871.

———. "The Weird, Wonderful World of Bioluminescence." Filmed March 2011 at TED2011, Long Beach, California. TED video, 12:30. www.ted.com/talks/edith _widder_the_weird_wonderful_world_of_bioluminescence?language=en#t -645745.

Widder, E. A., M. I. Latz, P. J. Herring, and J. F. Case. "Far Red Bioluminescence from Two Deep-Sea Fishes." *Science* 225, no. 4661 (September 1984): 512–14.

Chapter 12: The Edge of the Map

Bassler, Bonnie L. "How Bacteria Talk to Each Other: Regulation of Gene Expression by Quorum Sensing." *Current Opinion in Microbiology* 2, no. 6 (1999): 582–87.

Bracken-Grissom, H. D., E. Widder, S. Johnsen, C. Messing, and T. Frank. "Decapod Diversity Associated with Deep-Sea Octocorals in the Gulf of Mexico." *Crustaceana* 91, no. 10 (2018): 1267–75.

Czyz, A., B. Wrobel, and G. Wegrzyn. "*Vibrio harveyi* Bioluminescence Plays a Role in Stimulation of DNA Repair." *Microbiology* 146, no. 2 (February 2000): 283–88.

Fulton-Bennett, Kim. "Glow-in-the-Dark Corals Light Up the Deep Sea." Monterey Bay Aquarium Research Institute. Posted July 13, 2020. Accessed October 28, 2020. www.mbari.org/glowing-corals/.

Gaskill, M. "End of an Era for Research Subs." *Nature,* published online August 22, 2011. www.nature.com/news/2011/110822/full/news.2011.488.html.

Johnsen, S., T. M. Frank, S. H. D. Haddock, E. A. Widder, and C. G. Messing. "Light and Vision in the Deep-Sea Benthos: 1. Bioluminescence at 500–1000 m Depth in the Bahamian Islands." *Journal of Experimental Biology* 215 (2012): 3335–43.

Martini, S., V. Michotey, L. Casalot, P. Bonin, S. Guasco, M. Garel, and C. Tamburini. "Bacteria as Part of Bioluminescence Emission at the Deep ANTARES Station (North-Western Mediterranean Sea) During a One-Year Survey." *Deep-Sea Research Part I* 116 (October 2016): 33–40.

Messing, C. "Sea Star on a Stick: Introducing Crinoids." vimeo.com/410445885.

NOAA Ocean Explorer. "Bioluminescence 2009: Living Light on the Deep-Sea Floor." Accessed October 28, 2020. oceanexplorer.noaa.gov/explorations/ 09bioluminescence/welcome.html.

————. "Bioluminescence and Vision on the Deep Seafloor 2015." Accessed October 28, 2020. oceanexplorer.noaa.gov/explorations/15biolum/.

Robison, B. H., and K. R. Reisenbichler. "*Macropinna microstoma* and the Paradox of Its Tubular Eyes." *Copeia* 2008, no. 4 (December 2008): 780–84.

Tamburini, C., et al. "Deep-Sea Bioluminescence Blooms After Dense Water Formation at the Ocean Surface." *PLOS One* 8, no. 7 (2013): e67523.

Tanet, L., S. Martini, L. Casalot, and C. Tamburini. "Reviews and Syntheses: Bacterial Bioluminescence—Ecology and Impact in the Biological Carbon Pump." *Biogeosciences* 17, no. 14 (2020): 3757–78.

Thomas, Kate N., Bruce H. Robison, and Sönke Johnsen. "Two Eyes for Two Purposes: In Situ Evidence for Asymmetric Vision in the Cockeyed Squids *Histioteuthis heteropsis* and *Stigmatoteuthis dofleini*." *Philosophical Transactions of the Royal Society B: Biological Sciences* 372, no. 1717 (April 2017): 20160069.

Wagner, H. J., R. H. Douglas, T. M. Frank, N. W. Roberts, and J. C. Partridge. "A Novel Vertebrate Eye Using Both Refractive and Reflective Optics." *Current Biology* 19, no. 2 (January 2009): 108–14.

Warrant, E. J., and N. A. Locket. "Vision in the Deep Sea." *Biological Reviews* 79, no. 3 (2004): 671–712.

Widder, E. A. "Glowing Corpses and Radiant Excrement: The Role of Bioluminescence in Microbial Communities." In *Social Biology of Microbial Communities*, edited by LeighAnne Olsen et al., 533–45. Washington, D.C.: National Academies Press, 2012.

————. "Lighting the Deep." *New Scientist*, November 28, 2007, 24–25.

Zarubin, M., S. Belkin, M. Ionescu, and A. Genin. "Bacterial Bioluminescence as a Lure for Marine Zooplankton and Fish." *Proceedings of the National Academy of Sciences* 109, no. 3 (January 17, 2012): 853–57.

Chapter 13: The Kraken Revealed

Jabr, Ferris. "Gleaning the Gleam: A Deep-Sea Webcam Sheds Light on Bioluminescent Ocean Life." *Scientific American*, August 5, 2010.

Kiørboe, T. "How Zooplankton Feed: Mechanisms, Traits and Trade-Offs." *Biological Reviews* 86, no. 2 (May 2011): 311–39.

Kubodera, T., and K. Mori. "First-Ever Observations of a Live Giant Squid in the Wild." *Proceedings of the Royal Society B: Biological Sciences* 272, no. 1581 (September 2005): 2583–86.

Langlois, J. "How Scientists Got That Amazing Giant Squid Video." *National Geographic*, June 25, 2019. www.nationalgeographic.com/animals/2019/06/giant-squid-us-waters-first-video/.

McClain, C. R., A. P. Allen, D. P. Tittensor, and M. A. Rex. "Energetics of Life on the Deep Seafloor." *Proceedings of the National Academy of Sciences* 109, no. 38 (2012): 15366–71.

Nilsson, D.-E., E. J. Warrant, S. Johnsen, R. Hanlon, and N. Shashar. "A Unique Advantage for Giant Eyes in Giant Squid." *Current Biology* 22, no. 8 (April 24, 2012): 683–88.

NOAA Ocean Exploration and Research. "Journey into Midnight: Light and Life Below the Twilight Zone." Accessed October 28, 2020. https://oceanexplorer .noaa.gov/explorations/19biolum/welcome.html.

Widder, E. A. "Glowing Life in an Underwater World." Filmed April 2010 on Mission Blue Voyage, Galápagos Islands, Ecuador. TED video, 17:05. www.ted.com/ talks/edith_widder_glowing_life_in_an_underwater_world?language=en#t -695871.

———. "How We Found the Giant Squid." Filmed February 2013 at TED2013, Long Beach, California. TED video, 8:23. www.ted.com/talks/edith_widder_how_we_ found_the_giant_squid?language=en.

———. "The Kraken Revealed: The Technology Behind the First Video Recordings of Live Giant Squid In Situ." *Sea Technology*, August 2013, 49–54.

Wikipedia. S.v. "Deep-Sea Gigantism." Last modified October 28, 2020. en.wikipedia .org/w/index.php?title=Deep-sea_gigantism&oldid=982445652.

Chapter 14: Talking to Cannibals

American Museum of Natural History. "The Sorry Story of the Georges Bank." Accessed December 7, 2020. www.amnh.org/explore/videos/biodiversity/will-the -fish-return/the-sorry-story-of-georges-bank.

Broad, William J. "A New Ship's Mission: Let the Deep Sea Be Seen." *New York Times*, September 17, 2020. www.nytimes.com/2020/09/17/science/ocean -exploration-dalio-ship.html.

Burford, B. P., and B. H. Robison. "Bioluminescent Backlighting Illuminates the Complex Visual Signals of a Social Squid in the Deep Sea." *Proceedings of the National Academy of Sciences* 117, no. 15 (2020): 8524–31.

Galeazzo, G. A., et al. "Characterizing the Bioluminescence of the Humboldt Squid, *Dosidicus gigas* (d'Orbigny, 1835): One of the Largest Luminescent Animals in the World." *Photochemistry and Photobiology* 95, no. 5 (September/October 2019): 1179–85.

Gopnik, A. "Explanation as Orgasm and the Drive for Causal Understanding: The Evolution, Function and Phenomenology of the Theory-Formation System." In *Cognition and Explanation*, edited by F. Keil and R. Wilson, 299–323. Cambridge, Mass.: MIT Press, 2000.

Rosen, H., W. Gilly, L. Bell, K. Abernathy, and G. Marshall. "Chromogenic Behaviors of the Humboldt Squid (*Dosidicus gigas*) Studied In Situ with an Animal-Borne Video Package." *Journal of Experimental Biology* 218, no. 2 (2015): 265–75.

Taub, Ben. "Thirty-Six Thousand Feet Under the Sea." *New Yorker*, May 10, 2020. www.newyorker.com/magazine/2020/05/18/thirty-six-thousand-feet-under -the-sea.

Widder, E. A. "Making Light: Hunting for the Humboldt Squid with Edith Widder." OceanX Media. Posted January 26, 2018. Vimeo video, 3:22. vimeo.com/ 252948862.

Widder, E. A., C. H. Greene, and M. J. Youngbluth. "Bioluminescence of Sound-

Scattering Layers in the Gulf of Maine." *Journal of Plankton Research* 14, no. 11 (1992): 1607–24.

Wikipedia. S.v. "*Blue Planet II.*" Last modified July 4, 2020. en.wikipedia.org/w/index.php?title=Blue_Planet_II&oldid=966060427.

Epilogue: A Case for Optimism

Levin, L. A., et al. "Global Observing Needs in the Deep Ocean." *Frontiers in Marine Science* 6, no. 241 (May 2019). www.frontiersin.org/articles/10.3389/fmars.2019.00241/full.

Smithsonian Marine Station at Fort Pierce. "Education and Conservation Links, Indian River Lagoon Species Inventory." naturalhistory2.si.edu/smsfp/irlspec/IRL_Links.htm.

Spinrad, Richard W. "The New Blue Economy: A Vast Oceanic Frontier." *Eos,* June 8, 2016. doi.org/10.1029/2016EO053793.

Thosteson, E. D., E. A. Widder, C. A. Cimaglia, B. Burns, and J. Taylor. "New Technology for Ecosystem-Based Management: Marine Monitoring with the Kilroy Network." In *Oceans 2009 Europe,* 1–7. Piscataway, N.J.: Institute of Electrical and Electronics Engineers, 2009.

Widder, E. A. "A Pollution Map Worth Ten Thousand Words." *Open Science EU,* June 2016, 72–73.

Widder, E. A., and B. Falls. "Review of Bioluminescence for Engineers and Scientists in Biophotonics." Special issue, *IEEE Journal of Selected Topics in Quantum Electronics* 20, no. 2 (2013): 1–10.

Widder, E. A., B. Falls, R. R. Rohm, and C. Lloyd. "Save the Water Babies: High School Students as Citizen Scientists." *The Journal of Marine Education* 29, no. 1 (2014): 16–21.

Wikipedia. S.v. "James Stockdale." Last modified October 24, 2020. en.wikipedia.org/wiki/James_Stockdale.

INDEX

ABOUT THE AUTHOR

EDITH WIDDER is an oceanographer, a marine biologist, and the cofounder, CEO, and senior scientist at the Ocean Research & Conservation Association, a nonprofit organization where she is channeling her passion for saving the ocean into developing innovative technologies needed to preserve and protect the ocean's most precious real estate: its estuaries. She has given three TED Talks and been awarded a prestigious MacArthur "Genius Grant" from the John D. and Catherine T. MacArthur Foundation as well as the Explorers Club Citation of Merit, and is the first recipient of the Captain Don Walsh Award for Ocean Exploration, established by the Marine Technology Society and the Society for Underwater Technology.

teamorca.org